Manual of Water Supply Practices

M29

Water Utility Capital Financing

Fourth Edition

American Water Works Association

Manual of Water Supply Practices—M29, Fourth Edition

Water Utility Capital Financing

Disclaimer
The authors, contributors, editors, and publisher do not assume responsibility for the validity of the document or any consequences of its use. In no event will AWWA be liable for direct, indirect, special, incidental, or consequential damages arising out of the use of information presented in this book. In particular, AWWA will not be responsible for any costs, including, but not limited to, those incurred as a result of lost revenue. In no event shall AWWA's liability exceed the amount paid for the purchase of this book.

Sr. Manager, Editorial, Development and Production: Gay Porter DeNileon
Project Manager: Melissa Valentine
Senior Manuals Specialist: Molly Beach
Production: TIPS Technical Publishing, Inc.

If you find errors in this manual, please email books@awwa.org. Possible errata will be posted at www.awwa.org/resources-tools/resource.development.groups/manuals-program.aspx.

Library of Congress Cataloging-in-Publication Data

Fedder, Richard.
Water utility capital financing / Richard (Toby) Fedder, Eric Hofeld, John Mastracchio. -- Fourth edition.
pages cm -- (AWWA manual ; M29)
Revised edition of the Associations's Fundamentals of water utility capital financing, 3rd ed., published in 2007.
Includes index.
ISBN 978-1-62576-016-6
1. Municipal water supply--United States--Finance--Handbooks, manuals, etc. 2. Waterworks--United States--Finance--Handbooks, manuals, etc. 3. Water utilities--Finance--United States--Handbooks, manuals, etc. I. American Water Works Association. II. Fundamentals of water utility capital financing. III. Title.
HD4461.F43 2014
363.6'10681--dc23
2014030193

ISBN-13 978-1-58321-016-6
eISBN-13 978-1-61600-277-3

Printed in the United States of America
American Water Works Association
6666 West Quincy Avenue
Denver, CO 80235-3098
awwa.org

Contents

Figures

This page intentionally blank.

Tables

This page intentionally blank.

Preface

The American Water Works Association (AWWA) recognizes that customers' and the general public's interests are best served through well-maintained and efficiently operated water systems that are sufficiently funded to meet a utility's annual operating and capital needs. In support of this position, AWWA has published and regularly updated the M29 manual, now titled *Water Utility Capital Financing*, to assist utility management and stakeholders address the capital financing needs of their water systems. The terminology used in this manual is applicable to the United States and Canada; the manual does not attempt to identify practices and terminologies used in the myriad other countries with public utilities. It is likely that many countries outside the United States and Canada have financial instruments and government-sponsored programs that resemble those that are referenced in this manual. The financial management strategies presented herein are seen by AWWA as universal and thus valuable to all who make use of this document to guide their programs and decisions.

The financial operations of many water utilities are continually challenged by a variety of factors, such as adverse weather conditions, inflation, changes in the economy and/or political environment, growth or decline in service demand due to new construction, conservation programs, changes in environmental regulations, backlog of infrastructure repairs and improvements, staff turnover, and above all, public pressure to keep water rates low. Regardless of such challenges, water utilities remain very capital-intensive businesses. Despite the growing maintenance needs of aging water systems, many water utilities elect to postpone repairs, replacements, and rehabilitation expenditures to keep current operating costs, and corresponding service charges, as low as possible. However, deferring capital expenditures invariably results in higher operating costs in the short run and higher construction costs in the future.

Strategic planning helps utility management develop the steps necessary to reach the utility's short- and long-term goals and objectives. A strategic financial plan helps management meet the utility's operating and capital financing needs during a specified period. Similarly, strategic capital planning, while consistent with and supported by the strategic financial plan, addresses the utility's smaller recurring capital needs as well as its long-term expansion, upgrade, replacement, and reliability needs.

Strategic financial and capital plans also provide a platform for communicating the utility's needs to the public, thereby gaining greater stakeholder confidence in utility management. The plans inform investors of a utility's potential and support rating agencies' debt ratings. Whatever a particular water utility's capital requirements may be, knowledge of available alternatives for obtaining funding is a key element in developing successful financing plans. Some financing alternatives come into, or are phased out of, current practice with changes in federal tax law and market conditions. Thus, an effective capital financing program should contain a variety of fundamental approaches to financing infrastructure that create a systematic means of addressing the utility's needs over time. One resource that should not be overlooked is the utility's financial advisor, who is often able to propose current financing alternatives that may provide greater benefit to the utility than originally envisioned in a utility's financing plan.

PURPOSE

This manual provides a comprehensive overview of best management financing practices that water utility management should evaluate when determining short- and long-term strategies for meeting capital requirements. Only by thorough analysis of a particular water utility's long-term capital requirements and financial position will utility management be able to determine the feasibility of one or more specific financing concepts.

This manual is not intended to be, nor should it be considered, a complete text on water utility capital-financing concepts. A variety of other sources are available through various industry and professional organizations. Instead, this manual should primarily be considered a guide for determining capital requirements, strategic planning, and financing alternatives. It also provides general information for those utilities considering the issuance of long-term debt. Each of these areas is critical to the capital planning and financing process but must be tailored to and reflect factors applicable to a local situation. No recommendations or opinions about the relevance of alternative forms of available financing have been provided. Determining the relevance of alternative financing options is considered to be the prerogative of utility management.

ORGANIZATION

The material presented in this manual is intended for use by both government-owned and investor-owned water utilities. Because financing alternatives and markets may differ for these two primary types of ownership, financing alternatives for each are discussed in separate places throughout this manual.

This manual covers six general areas:

1. The capital and financial planning process
2. Identification of financing alternatives
3. Evaluation of financing alternatives
4. The process of taking a debt issue to market
5. Participants' roles in the debt issuance process
6. Special considerations for investor-owned utilities

Information presented in this manual has certain limitations. Water utility capital requirements continually change, as do financial sources available to the utility. At any given time, financing alternatives available to a utility depend on:

- Tax laws and Internal Revenue Service rules and regulations
- Interest rates charged for various financing alternatives

- A utility's credit rating
- Decisions by governing bodies, by regulatory agencies, and in some instances, by investors, ratepayers, and/or voters

Appendix A provides a summary of capital project prioritization using asset management principles. Appendix B provides example financial plans based on various capital financing scenarios. Appendix C provides information regarding the general requirements of bond-rating agencies, which is particularly useful as a utility considers the use of long-term debt and begins the process of going to the financial markets. The glossary provides terms commonly used by the financial community and is a continuing step toward establishing uniform definitions.

AWWA advocates that specific financing requirements for any particular water utility be based on sound economic, financial, accounting, and engineering principles. Often the services of consultants and counselors experienced in such matters are required. When such advisors are employed, this document should serve as a resource that the policymakers and utility managers may draw on to guide their evaluations of the validity of any proposed financing alternative.

Several other AWWA manuals and books address financial management topics that may interest the reader are identified throughout the text and in the reference and bibliography sections.

This page intentionally blank.

Acknowledgments

The AWWA Management and Leadership Division gratefully acknowledges the contributions made by those volunteers who drafted, edited, and provided the significant and critical commentary essential to updating M29. The Editorial Review Board members dedicated countless hours in the final stages of preparation of this edition to ensure the overall technical quality, consistency, and accuracy of the manual.

Editorial Review Board Members

Richard (Toby) Fedder, Woodard & Curran, Andover, Mass.
Eric Hofeld, Retired, Portland Water, Portland, Ore.
John Mastracchio, Arcadis U.S., Inc., Clifton Park, N.Y.
Margaret Medellin, AWWA, Denver, Colo.
Kenneth Mercer, AWWA, Denver, Colo.
James (Mike) Mussman, The Mussman Group, Frederick, Md.

Contributors to the fourth edition

Angela Bricmont, Denver Water, Denver, Colo.
Derrick J. Brown, Retired, Greenville Water, Greenville, S.C.
Ann Casey, CDM Smith, Kansas City, Mo.
Edward Donahue, Municipal & Financial Services Group, Annapolis, Md.
Richard (Toby) Fedder, Woodard & Curran, Andover, Mass.
Moira Geer, City of Winnipeg Water & Waste Department, Winnipeg, Man., Canada
Eric Hofeld, Retired, Portland Water, Portland, Ore.
Darin Kamradt, Water District No. 1 of Johnson County, Lenexa, Kan.
Bryan Mantz, Public Resources Management Group Inc., Maitland, Fla.
John Mastracchio, Arcadis U.S., Inc., Clifton Park, N.Y.
James (Mike) Mussman, The Mussman Group, Frederick, Md.
Robert Ori, Public Resources Mgmt. Group Inc., Maitland, Fla.
Courtney Rogers, Davenport & Company LLC, Richmond, Va.
Robert Ryall, Wildan Financial Services, Orlando, Fla.
James Sanderson, Davenport & Company LLC, Richmond, Va.
Lucy Szkwarek, City of Winnipeg, Finance Division, Winnipeg, Man., Canada
John Wright, Raftelis Financial Consultants, Centennial, Colo.

This page intentionally blank.

Chapter 1

The Capital Financial Planning Process

The demands and expectations of water utility stakeholders (i.e., customers, regulators, and policymakers) create a challenging capital financing environment for water utilities. In addition to demanding consistently safe and reliable drinking water, stakeholders expect water service to be provided efficiently at a reasonable cost, and with an appropriate burden to ratepayers.

This chapter provides an overview of the capital financial planning process. The discussion emphasizes the importance of assessing current and future capital requirements and evaluating the common assumptions used in capital program development. Water utility managers should perform this evaluation as they move from reviewing historical capital programs and assessing past financial performance to taking the specific strategic, financial planning, and budgeting action steps necessary to successfully finance the construction of current and future capital projects.

Developing a comprehensive capital financial plan requires more than life-cycle costing for individual capital projects. The financial impacts on utility operations and customer bills must also be considered. Accordingly, a capital financing plan, including the revenue requirement and rate impacts of alternative capital financing strategies, should be integrated into the utility's overall long-term financial planning process.

STRATEGIC PLANNING—THE FIRST STEP

Strategic planning is a dynamic and continual process providing direction for operations as well as utility infrastructure additions and replacements. The strategic plan should describe when and how resources will be obtained and used as well as define outcomes and outputs that facilitate performance monitoring. It also serves to guide the development of detailed policies that impact resource allocation and operational decisions.

Long-term strategic planning requires a long-term look at the future operations of a utility and maps a path between a utility's present position and its vision for the future. It assesses current conditions, strengths and weaknesses, identifies underlying causes and effects, and establishes a direction for accomplishing the utility's vision and objectives.

An effective strategic plan communicates management's goals for the future to customers, employees, rating agencies, investors, and the community as a whole. When developed in concert with an asset master plan, it also provides a framework for developing multiyear financial and capital improvement plans that can be implemented via annual budgets and spending plans. Figure 1-1 shows the steps used in developing a strategic plan with each step guided by the previous one.

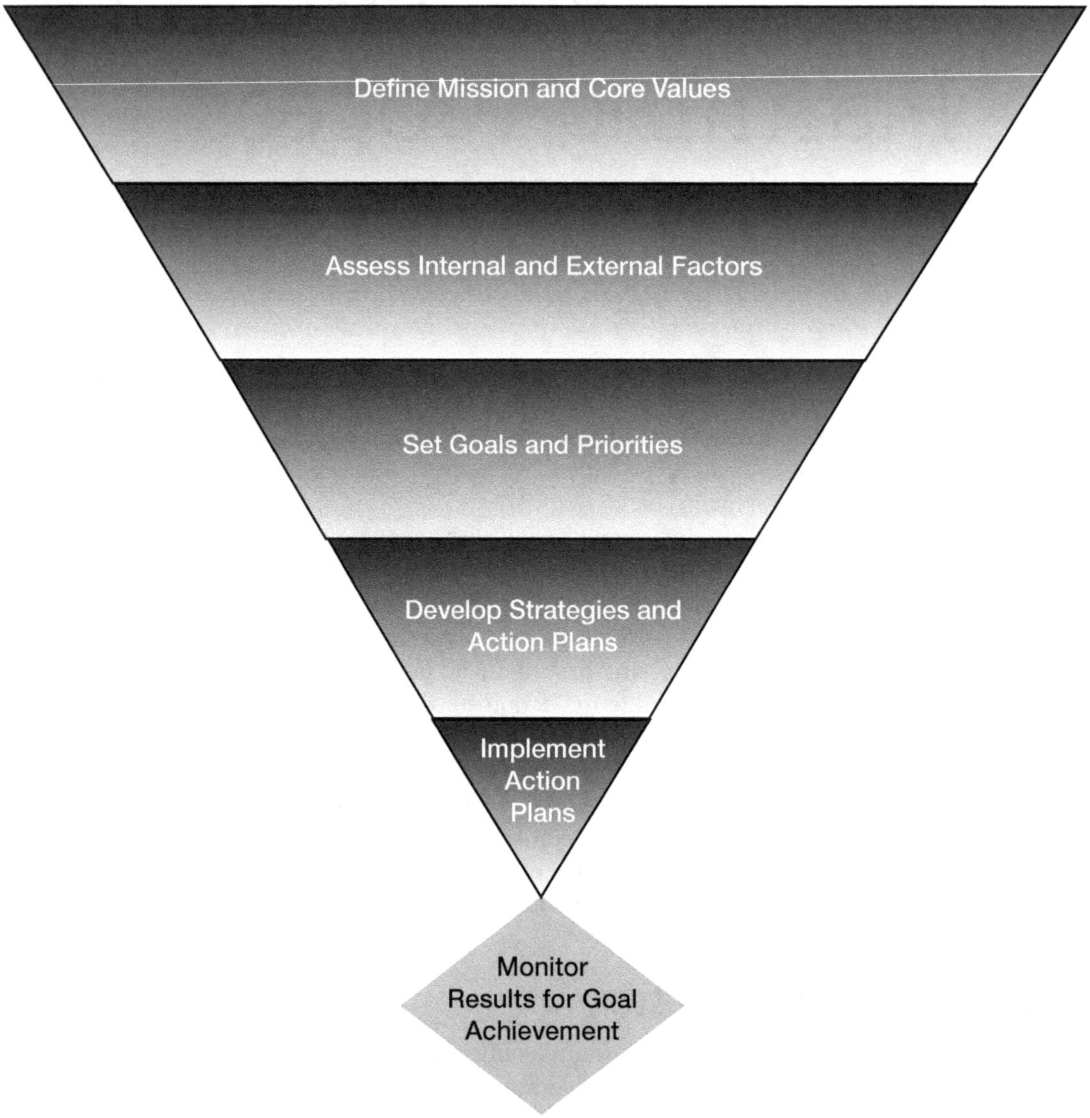

Figure 1-1 Strategic planning steps

Strategic Plan Components

As shown in Figure 1-1, the following components should be developed and/or incorporated into the strategic plan:

- Mission and core values
- Internal and external factors
- Goals and objectives
- Strategies and action plans

The *mission statement* describes a utility's basic reason for existence. This may be stated in legislation, corporate documents, or local ordinance.

Core values define how a utility will conduct its business and interactions with employees, ratepayers, and other stakeholders.

Internal and external factors should be listed and evaluated because of their potential to either help or hinder goal achievement. Internal factors are unique to the utility itself and include, but are not limited to, key staff and organizational structure, utility culture and image, utility plant condition, financial condition and strength, employee experience and turnover, operational efficiency, and technology.

External factors can be divided between those that affect many utilities and those that are unique to each water utility. External factors that affect many utilities include such things as changes in weather patterns and national and state/provincial economic conditions. External factors unique to the water utility include supply source and service area, changes in population and customer demographics, service-area economic conditions, customer water usage, federal and state regulatory mandates, and political climate. Each of these external factors affects long-term demand and the capital investments that will be needed.

Goals and objectives should be aligned with customer and stakeholder expectations, prioritized, and have clearly defined and measurable outcomes and outputs for evaluating success.

Finally, *strategies and action plans* for goal achievement should be developed. These are typically represented by separate multiyear capital improvement and multiyear financial plans based on the utility's asset management program and related infrastructure master plan. Figure 1-2 shows how each of these components are incorporated into developing a financial plan.

EVALUATE INTERNAL FACTORS

While many of the internal factors are important, those dealing with utility management and staff are beyond the scope of this manual. The two internal factors dealing with plant condition and financial condition and strength are factors most closely related to the capital financing process.

Risk Assessment and the Asset Management Program

The magnitude of expenditures that are or will be needed in the future to repair or replace aging water utility infrastructure has been well documented by the American Water Works Association (AWWA). Such expenditures are often a key component of a utility's capital requirements and can pose significant financial and rate affordability challenges. To meet these challenges in the most effective manner, utility managers need to ensure that capital expenditures for which they seek funding have been developed through a formal

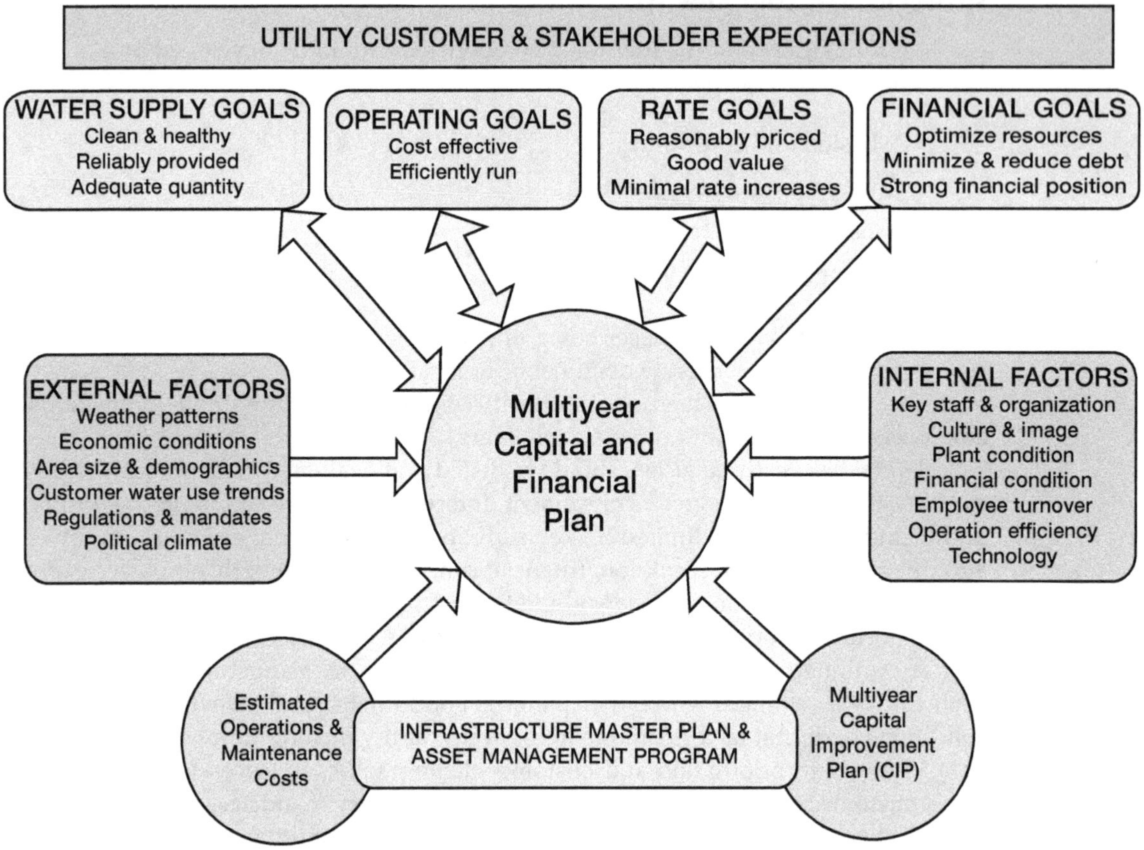

Figure 1-2 Capital and financial plan development

asset management program. A formal asset management program combines inputs from a variety of sources, including financial reports, asset operations and maintenance (O&M) records, geographic information, and risk assessment of utility plant assets.

Utility-plant asset risk assessment combines the likelihood of asset failure with the consequences of failure to the utility. When successfully implemented, asset risk assessment helps the utility develop a prioritized list of capital expenditures that are inputs to the infrastructure master plan and provide support for the multiyear capital improvement plan and the multiyear financial plan. A risk assessment also provides the utility with quantitative explanations to customers and stakeholders for planned expenditures that had previously been justified on a purely qualitative basis or after the occurrence of an actual asset failure.

A common practice in risk assessment is to assign a rating to each asset that combines the likelihood of failure with the consequence to the utility if the asset fails. For example, using a scale of 1 to 5, a 25-year-old pump in poor condition may have a probability of failure within the year of 5, but because it is used for decorative fountains, may only have a consequence of 1. So its combined rating is 5 (5 × 1). However, if a pump with the same probability of failure refills a major reservoir that serves a large portion of the utility's customers, the consequence of failure to the utility may be 5, producing a combined rating of 25 (5 × 5). By assigning each major asset a combined rating, a risk profile of water system's assets emerges, providing a way of prioritizing maintenance and/or replacement expenditures.

Table 1-1 is an example of a risk assessment matrix that could be used in asset maintenance/replacement planning. Capital projects that fall within the range of the highest likelihood of failure and the highest consequence of failure should receive the highest priority in terms of funding. Appendix A explains how risk assessment can be developed and used with an asset management program.

Infrastructure Master Plan

An infrastructure master plan may cover 20 or more years and has many benefits. It incorporates the utility's service goals and objectives (and stakeholder expectations). It also includes long- and short-term estimates of water service capacity needs, selection and design of system infrastructure, long-term capital needs for required infrastructure facilities, and when combined with asset risk assessments, the timely replacement and retirement of system assets. A master plan

- incorporates service goals and objectives,
- describes and characterizes the existing system,
- anticipates long-term capital needs, and
- guides prioritization of capital projects.

Utility managers should consider conducting a project-specific economic analysis when developing an infrastructure master plan, as well as a multiyear capital improvement plan. This analysis involves calculating the net present value and life-cycle cost of a capital project. It is often used to determine the most cost-effective project when more than one alternative is under consideration. This technique can be integrated with the asset management and master planning processes as part of a comprehensive approach to capital planning.

EVALUATE EXTERNAL FACTORS

External factors applicable to most utilities should be evaluated for their long-term impacts on the geographical region in which the utility operates. However, their impact on a utility's strategic plan may be much harder to quantify.

External factors unique to each utility will often have a more direct impact on the capital planning and financing process. Some, such as regulatory mandates or political climate, may be less predictable than those that impact future water demand. However, forecasting customer water demand, on both an annual and seasonal peak load basis, is important to determining a utility's future capital requirements.

Table 1-1 Risk assessment matrix

		Consequence of Failure				
		1	2	3	4	5
Likelihood of Failure	1	Very low	Very Low	Low	Low	Medium
	2	Very low	Low	Medium	Medium	High
	3	Low	Medium	Medium	High	Extreme
	4	Low	Medium	High	Extreme	Extreme
	5	Medium	High	Extreme	Extreme	Extreme

Long-Term Water Demand

Components that influence long-term water demand are changes in weather patterns, service territory, population, consumption (per capita, account, or unit), customer demographics, and land-use patterns.

Population forecasts can often be obtained from a variety of federal, state/provincial, or local government agencies. Compare forecasts made by external parties against actual utility experience during the previous 5- to 10-year period for the most accurate picture. Through the thoughtful use of both external and internal data, the utility can develop a long-term forecast of service area population and associated customer account growth suitable for demand forecasting and financial planning purposes.

Changing customer class demographics and changing land-use patterns also can have a large impact on water demand. For example, a shift in the profile of the commercial customer class to—or away from—large commercial and industrial customers to smaller service-oriented industries can greatly influence long-term demand. Changing land-use patterns, such as the movement from large-lot single family residential construction to higher density multi-family housing, can change both the amount and intensity of annual and peak-load water demands.

After completing a forecast of customer growth, water demand forecasts can be developed for each customer class served by the utility. To accomplish this, the utility may elect to undertake a comprehensive analysis of the historical trends in customer demand, on an average consumption-per-account-class basis. Such an analysis can be useful because many utilities throughout North America have experienced steady declines in the average residential consumption, and in some cases, declines in commercial customer consumption. Understanding the reason for these declines and using the most updated assumptions regarding customer consumption will facilitate more accurate long-term demand forecasts as part of the financial planning process.

Demand Forecasting Scenarios

For planning purposes, a utility may want to use regression analysis or Monte Carlo simulations to develop a range of long-term demand forecasting scenarios. Monte Carlo simulations rely on repeated random sampling to obtain a distribution of numerical results. The range is a characterization of probability based outcomes that provides intervals around expected values, allowing the user to select a scenario for planning purposes based on a utility's aversion to risk. A scenario that reflects the utility's most acceptable risk is then selected and used for conducting the remaining planning process.

Demand scenarios can be interchangeable according to the priority of the goals used to develop a utility's strategic plan. For example, if new supply is a high priority because the utility does not have the capacity to serve additional demand, one scenario is a rapid increase in forecasted demand. This scenario may require immediate and costly capital investments to increase supply, treatment, transmission, and distribution infrastructure.

A higher risk-based scenario would be to assume a forecast of lower demand and slower growth, with the lowest demand simulations creating the highest risk of not having sufficient capacity. However, if supply is not a high priority, and the utility has excess capacity, a worst-case scenario would be a declining customer base. An appropriate scenario to plan for is one where risks and predictions are balanced and correspond with the risk tolerance of the organization.

Depending on their intended use, alternative scenarios may be required. For example, demand projections suitable for facility planning may result in higher risk for financial planning purposes. The goal of facility planning is to ensure that service demands can be met. Hence, demand projections that assume increased population growth (worst case) might be appropriate for facility-planning purposes. However, basing rates and revenue projections on the same growth estimates may understate required future rate revenues and increase financial risk for the utility. The timing of future rate revenues resulting from projected customer usage is a key financial planning element. Frequently, the use of two scenarios—one for facility planning and a second for financial planning—may be appropriate.

The acid test for any scenario is the utility's ability to demonstrate the probability that the chosen scenario will occur. In making the final presentation for budget approval, only the most probable scenario should be chosen for planning purposes. The chosen scenario should produce results that represent the least financial risk to the utility and ratepayers while ensuring that properly sized plant additions will be placed in service before capacity is required to meet increased demand.

Figure 1-3 illustrates the trade-offs between reliability and cost and indicates that utilities strive to strike a balance between the amount of system reliability and the cost of implementing and maintaining a reliable system. Figure 1-4 illustrates two possible cases for the same level of reliability and indicates that reliability could be measured by the frequency of shortages occurring as well as the magnitude and impact of the shortage when it occurs. In any case, the planning process should result in a common agreement by both the utility and its stakeholders and ratepayers as to what are acceptable levels of reliability and cost.

Integrated Resource Planning

One technique used by many utilities to identify the optimal portfolio of new supply side resources and demand management programs is integrated resource planning (IRP). IRP has its roots in a method of planning used by electric and gas utilities in the 1980s when supply was short and the cost of additional supply became a significant factor. IRP brings an important dimension into the strategic planning process for water utilities and opens the door to explicit evaluation of options that may not have been previously considered,

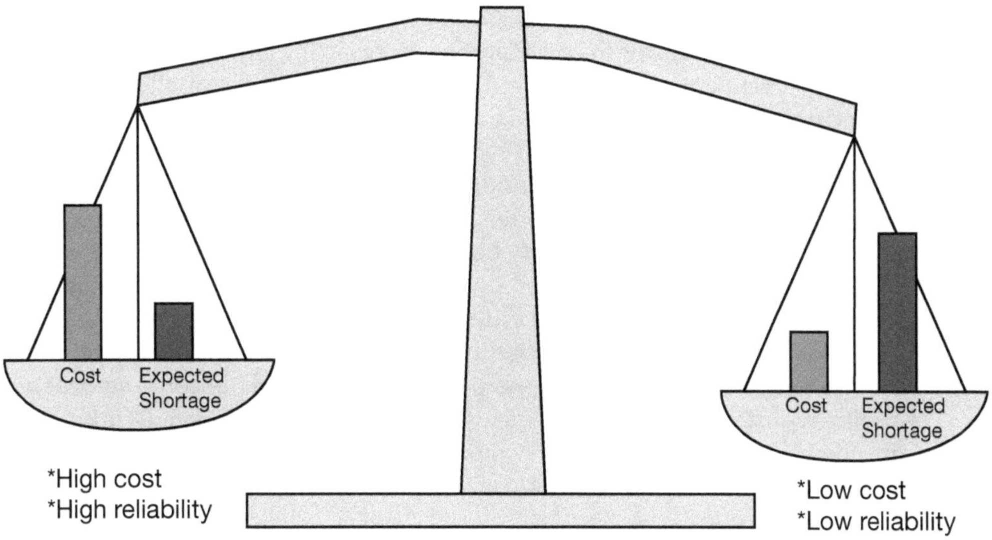

Figure 1-3 Trade-off between reliability and cost

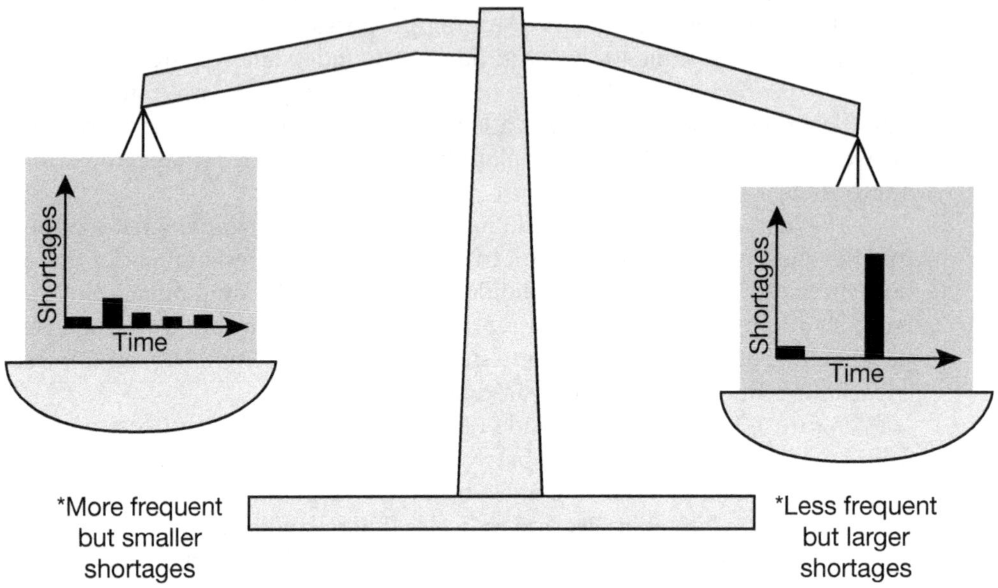

Figure 1-4 Different aspects of reliability

such as wastewater reuse, demand management, conservation, and planned shortages all with expanded stakeholder involvement.

Basically, strategic planning using IRP looks not only at cost and availability of supply but also at a utility's ability to manage demand through conservation and/or public outreach efforts. The process provides information on the consequences of various options and helps management judge the value of trade-offs among resource alternatives. When properly applied, the IRP process leads to sound long-term decisions and lower overall costs. A more detailed explanation of the IRP process can be found in AWWA's Manual of Water Supply Practices M50, *Water Resources Planning*.

Utility Regulation

Another important factor to consider when forecasting utility capital requirements is the effect of current and proposed regulations and the dictates of all regulatory agencies at the federal, state/provincial, and local levels. For investor-owned water utilities, regulatory control is usually exercised by a state or provincial regulatory agency, such as a public utilities commission. The authority given these agencies is generally comprehensive. Government-owned utilities are often controlled and regulated by local governments such as city councils or municipal utility boards. In some states/provinces, government-owned utilities serving customers located outside their municipal boundaries are regulated by state or provincial agencies. Such utilities need to take steps to ensure they are in compliance with these regulatory agencies and verify that allowances in capital and operating budgets have been made to address deficiencies that may become evident as a result of specific regulations. For example, legislation such as the US Safe Drinking Water Act, as amended, and subsequent US Environmental Protection Agency regulations, which affect how all US utilities operate, can cause many utilities to make major modifications to their capital plans.

SET GOALS AND PRIORITIES

After considering all of the relevant issues, goals should be developed that address each of the relevant internal and external factors. In this process, some goals may conflict with others so the impact of each factor must be considered in order to prioritize goals for optimum resource allocation. Lower priority goals that conflict with higher priority goals should receive little or no resources.

Prioritized goals then become part of the utility's multiyear capital and financial plans. These plans should include comprehensive yet concise statements that consider and address each relevant factor.

Targets for specific action steps can be set during the planning phase; however, the results of those actions should be continually measured during the implementation phase by asking the following questions:

- Are we doing what we planned to do?
- Are we achieving the intended results?

The utility should conduct periodic assessments that measure both effectiveness and efficiency in reaching stated goals. A good assessment process answers both of these questions. The Governmental Accounting Standards Board (GASB) also encourages governmental entities to experiment with performance measures and standards reporting and to integrate performance measures into a utility's capital and operating programs.

Both the multiyear capital and financial plans must be fluid and will need to be updated as changes occur or when action steps are not yielding the desired results.

THE MULTIYEAR CAPITAL IMPROVEMENT PLAN

The multiyear capital improvement plan (CIP) is typically a 5- to 10-year forecast of infrastructure facility plans with estimated capital spending. Utilities may use longer planning periods; for example, from the subsequent fiscal year to service area build-out; however, the longer the planning window, the less reliable the estimates tend to be.

The multiyear CIP should reflect management's near-term goals and priorities and risk assessments that often include major as well as routine capital projects. Major capital projects include proposed additions and improvements that involve relatively large expenditures and are nonrecurring in nature. Routine capital projects include proposed additions, replacements, and improvements involving relatively small expenditures that may occur annually.

Developing the Multiyear CIP

The process of developing a multiyear CIP should integrate risk assessments and other aspects of planning. Include both the priorities for existing and new facilities and the impacts on operating and capital costs and associated rate revenues. The effect of the CIP on personnel and material resources serves as the basis for planning the operational and construction activities the utility will perform. Remaining activities may have to be outsourced. Through this comprehensive evaluation process, the plan provides an instrument for control over capital expenditures. The multiyear CIP should be reviewed and revised annually. As the current year drops off, plans for the next several years are reviewed and modified and the CIP extended for at least one additional year.

CIP Implementation

One of the most common indicators of strategic planning success is a utility's ability to implement its adopted CIP. A utility develops a schedule for completion of CIP projects based on goals, internal and external factors, and the infrastructure master plan. The schedule becomes a fundamental part of the utility's CIP. Table 1-2 shows a sample CIP schedule of project priorities resulting from a utility's capital planning process. An adopted CIP has a schedule and budgeted cost based on estimates, as described in later sections of this chapter. The CIP provides basic guidance for determining when each project will be implemented and how much of the utility's resources can be allocated toward completion of projects at any given time. Success in implementing the CIP is measured by comparing budgeted capital projects to actual work completed and actual costs incurred to complete work during the budget period.

Ideally, the CIP is completed as scheduled, demonstrating the utility's ability to achieve its stated goals. However, there are numerous reasons why a CIP schedule might not be met. The causes should be identified and evaluated. If schedule slippage is within management's control, action should be taken to correct identified problems during the planning process or as detailed action plans are completed. However, slippage may also occur because of changes in one or more factors or because the assumptions used to develop the strategic plan have changed. Regardless of the reason for falling behind, the utility needs to adjust its multiyear CIP and financial plan to reflect current implementation capabilities and changes in its allocation of resources.

Table 1-2 Sample CIP schedule

		Estimated Cost ($ millions)				
Project Name	Risk Score*	FY 2013	FY 2014	FY 2015	FY 2016	FY 2017
Treatment facilities improvements	25	40.00	19.00	3.00		
Terminal reservoirs	25	78.05	65.50	18.25	12.00	20.50
Wellfield reliability enhancements	20	0.04	0.13	1.96		
Transmission mains	12	0.35		1.00	9.00	5.00
Distribution mains	12	6.60	6.60	6.60	6.60	6.60
Pump station improvements	12	5.00	3.15	0.20	1.00	2.00
ESA compliance—capital outlay	10	15.40	1.20	1.00	9.00	2.00
Interstate facilities rehabilitation	10	10.00	10.00	20.00	1.00	
Willamette River pipe crossing	10			2.60	5.00	20.00
Tank improvements	9	4.00	0.50	0.50	0.50	1.00
Conduit repair and rehabilitation	9			0.50	1.00	1.00
Services	8	3.50	3.50	3.20	3.50	3.50
Utility line relocates	8	2.00	0.10	0.10	0.10	0.10
Distribution mains	6				2.50	2.50
Hydrant replacement	6	0.50	0.80	0.80	0.80	0.80
Meter purchases	6	1.59	1.59	1.59	1.59	1.59
SCADA improvements	5	0.28	0.28	0.28	0.28	0.28
Wholesale connections	5		0.025	0.30	0.80	0.10
Conduits transmission	4				5.0	5.0
Equipment purchases	4	3.41	3.41	3.41	3.41	3.41
Fountain maintenance	3	0.15	0.15	0.15	0.15	0.15
Capital planning studies	3	1.10	1.10	1.10	1.10	1.10
Totals	172.27	117.035	66.84	66.54	64.33	76.63

* The risk score is the combination of score for the likelihood of failure with score for the consequence of failure. Projects with high probability of failure and high consequence of failure receive the highest risk score and should be highest in priority for funding.

THE MULTIYEAR FINANCIAL PLAN

The multiyear CIP guides the utility in carrying out the strategies included in its infrastructure master plan by providing a "road map" for capital improvements. Examples of multiyear capital financing scenarios are presented in appendix B. To implement this road map, the utility should follow these steps to develop a multiyear financial plan:

1. Assess current financial condition.
2. Estimate planning period's capital and operating costs.
3. Identify capital financing alternatives.
4. Calculate the amount of revenue needed to cover costs.
5. Assess rate and service impacts on customers.
6. Revise capital program and operating costs to stay within financial capability.

This financial plan should also describe the sources of funds to initially construct facilities and then to operate and maintain them after construction is complete.

Assess Current Financial Position

The first step in developing a multiyear financial plan is to assess the utility's current financial position. This step will identify available financial resources that may be used to finance the capital program, such as existing unrestricted cash reserves, debt authorizations, developer charges, and future revenues, and identify existing burdens on the utility, including annual debt service.

To assess current financial position, a utility must have accurate, detailed, and timely data. Well-managed utilities maintain good financial information systems. Such a system begins with a detailed accounting system that tracks assets, liabilities, revenues, and expenses over time and summarizes the results so that the current status of the utility can be evaluated and changes can be made in the future. Consult AWWA's handbook *Financial Management for Water Utilities* for more in-depth discussions on financial accounting, reporting, and analysis, as well as a detailed explanation about what is considered a capital expenditure.

By using information contained in standard financial reports, a utility can determine financial position and calculate key financial indicators that assess its ability to support additional debt for capital improvements. Financial rating firms such as Fitch Ratings (www.fitchratings.com), Standard & Poor's (www.standardandpoors.com), Moody's (www.moodys.com) and DBRS (www.dbrs.com) all provide indicators and industry statistics that can be used for such analysis (rating agencies are discussed in chapter 4).

In addition to financial information, a utility needs information on system use and customer accounting, billing, and collections. Comparing annual water bills per household to median household income may indicate whether or not customers can reasonably afford and support higher rates for a new or expanding capital program.

To determine a community's ability to fund a capital program, financial analysts and rating agencies routinely measure financial performance. They rely on a compilation of financial, debt, socioeconomic, and management indicators, comparing them to regional and national averages to assess a community's ability to repay debt. By accurately determining its current financial position, a utility is better able to assess the impacts of a proposed CIP on stakeholders and customers.

Forecast Revenues and Costs

Financial planning requires projections of utility revenues and costs over the financial planning period and possibly for a longer period in the case of an infrastructure master plan. Although projections for periods longer than five years become increasingly speculative, they provide insight into long-term impacts of current business decisions. Economic and regulatory conditions affecting future service requirements are uncertain, but long-term capital planning requires that the effects of these factors on revenues and costs be included for the specified planning period.

Cost projections show proposed changes in

- capital spending as a result of risk assessments and/or anticipated changes in service requirements;
- reserve policies and/or debt coverage requirements;
- employee benefits and pensions, inflation, or operating trends; and
- revenue based on the number of customers served or anticipated rate adjustments.

Generally, cost projections are based on historical trends adjusted for any known changes in future conditions. Several sources can be consulted for price increases and related cost trends, such as the US Congressional Budget Office, the US Department of Labor's Bureau of Labor Statistics Consumer Price Index, (which includes data on water, sewer, and trash collection services combined), and the Producer Price Index. Construction cost indices (e.g., the Handy-Whitman Index of Public Utility Construction Costs, the *Engineering News-Record* Construction Cost Index) and municipal bond interest rates may also be used to project current and anticipated new costs into the future.

Projections over the planning period generally start with the annual budget. Projections of revenues, O&M costs, and capital expenditure are extrapolated over the planning period and adjusted for inflation, changes in service levels, and other anticipated changes in conditions. Specific project-related O&M expenses should also be included (e.g., additional operating personnel to staff a planned new water treatment plant).

An asset management program will allow the utility to track original cost, age, and accumulated depreciation of capital assets by function. Maintenance frequencies and costs also assist in risk assessment and in deciding whether to repair or replace capital assets. Note: A thorough financial analysis should prevent the unfortunate situation of a utility being unable to fund its CIP or worse, implementing a CIP only to discover later that the resources to repay construction-related debt, or to properly operate and maintain the assets after they are built, are not available.

After determining its capital requirements, a utility should identify the financing alternative that best suits its needs and resources. This process is discussed more fully in chapter 2.

Who Should Pay for Plant Additions?

A utility has an obligation to serve all customers, both existing and new. However, the capital cost of providing service to new customers can be significantly higher than the cost of serving existing customers. The utility should develop a policy that addresses cost recovery for growth-related additions to capacity (i.e., a determination of who pays for the cost of growth) for use in developing multiyear CIP and financial plans.

The answer to the question of who pays is not as clear as would first appear. Not all additions to capacity benefit only new customers. Some plant additions may also improve the quality of service to existing customers. Service objectives for existing and new customers are in many cases the same (water quality and safety, as well as service reliability, for example). However, current customers seek to minimize the impact of growth on general service charges, while new customers would like to minimize front-end costs to connect to the water system.

To the extent that capital additions satisfy similar objectives for both new and existing customers, the question of who pays becomes uncertain. However, capital costs that clearly benefit only new customers can be funded by those new customers directly through project-specific impact fees or system development charges. Capital costs can also be recovered from new customers connecting to the system through a reimbursement impact fee, which is a fee paid after the construction of the facilities. In some cases, utilities assess both types of charges depending on state/provincial law and the nature of the capital costs incurred. For a more in-depth discussion, refer to AWWA Manual M1, *Principles of Water Rates, Fees, and Charges*.

Calculate Revenue Needed to Cover Costs

Once financing alternatives are selected, the resulting cash flows, as well as related operating expenses, must be integrated into the multiyear financial plan and finally, into annual operating and capital budgets. Revenues by source are typically identified in both the multiyear financial plan and annual budget. These sources may include user charges, tax revenues, impact fees or system development charges, connection fees, and bond proceeds.

AWWA Manual M1 strongly supports the Association's policy statement regarding financing and rates, which stresses that the public is best served by enterprises that are adequately financed through rates based on sound engineering and economic principles and that do not require subsidies from other entities.

Assess Rate and Service Impacts on Customers

If additional revenues or spending reductions are needed to meet the utility's strategic service goals and objectives, they should be identified early in the process. The utility will need to determine whether the proposed level of spending will impose an undue financial burden on system users. If rate increases associated with a particular alternative are perceived to be too large, the planning process will need to be refined to either reduce project costs, adjust project schedules, identify alternative long-term financing, or find different sources of revenue.

Sensitivity to rate impacts often comes to the forefront in the development of optimal financial plans. Table 1-3 is a sample list of questions that should be considered during the financial planning process and the development of an optimum capital financing plan.

A key factor in obtaining the rate increases necessary to adequately fund a capital program is how much effort the utility puts into educating key stakeholders on the true value of water. Utilities must reset the common (mis)understanding of the value of water so that it gets the attention—and money—it deserves. This will take a two-pronged approach. Phase 1 is to win over decision makers using a rigorous asset management program. However, elected officials may be more motivated by votes than by what is "right" and raising water rates may not be as politically expedient. That's where the utility should also implement Phase 2—win over the public via community outreach. If the public truly understands and appreciates the value of water and, by extension, the role of the water utility, they are more likely to support the investment required to adequately fund the repair, expansion, and upkeep of vital water systems. Once the public is won over, governing bodies can grant approval on water rate increases with less fear of political fallout. Community outreach should be conducted in concert with asset management as a lead-up to funding proposals.

Table 1-3 Financial planning checklist

Yes	No	Assess Financial Condition
		Does the current financial condition meet or exceed industry standards?
		Do key financial ratios indicate the utility can support additional debt?
		Does the current credit rating allow access to the market at a reasonable cost?
		Can outstanding debt be defeased (refinanced) to save interest costs?
		Does the utility have excess bonding capacity?
		Are reserves adequate to smooth rate increases and meet debt coverage?
Yes	**No**	**Estimate Capital and Operating Costs**
		Is billed consumption trending upward or downward in all customer classes?
		Are there unintended negative impacts on the local economy?
		Are annual water bills growing faster than median household income?
		Has the public expressed concern over its ability to pay?
		Are operating costs growing faster than inflation?
		Does the utility have acceptable staff levels and expertise?
		Are employee benefits and pensions adequately funded?
Yes	**No**	**Identify Capital Financing Alternatives**
		Are other revenues available to fund capital and/or operating costs?
		Has the issue of "who pays for new capacity?" been resolved?
		Have alternative financing sources been identified and evaluated?
		Are capital cost estimates and implementation schedules reasonable?
		Have the utility's needs been fully and continually explained to customers?
		Have customers accepted the capital program and related rate increases?
		Have capital improvement costs and/or schedules been adjusted?

For more information about raising the public's and elected officials' awareness of the cost of providing water services, see the AWWA handbook *Communicating Water's Value: Talking Points, Tips & Strategies*, by Melanie Goetz.

Revise CIP to Stay Within Financial Capability

As previously discussed, evaluations of the financial impact of estimated future conditions should be based on multiyear analyses. The long-term impacts of bonding capacity, rate levels, and affordability for individual customers should also be considered when planning major projects. Infrastructure risk assessment needs to be considered because significant risk may exist even when planning for only one or two years. When extending capital planning over a longer period, it is imperative to minimize the possibility of making poor decisions that will have adverse consequences long-term.

Several measures may be used to evaluate financing alternatives, such as return on capital, internal rate of return, net present value, and life-cycle cost. Chapter 3 goes into this topic in more detail.

THE ANNUAL CAPITAL BUDGET

An annual capital budget is used to implement the CIP. An annual budget presents a picture of the course the utility is taking over the next fiscal year. It identifies what operating and capital expenditures will be incurred, including routine annual capital improvements and capital cost recovery components during the first year of the multiyear CIP.

For the CIP to be useful as a planning and control document, total anticipated capital expenditures should be supported by individual project budgets. Normally, project budgets specify a project number, brief project description, purpose and necessity for the project, the project's priority, the scheduled starting and estimated completion dates, a detailed cost estimate, an estimate of incremental operating revenues and expenses associated with the project, and, if applicable, the estimated return on investment. Although it may be difficult at times to trace the actual source of funds for individual projects, it is usually desirable to identify the intended financing source(s). This helps identify special financing opportunities, which may reduce the utility's total funding requirement.

Additional guidance on developing an annual capital budget can be found in AWWA's *Financial Management for Water Utilities.*

BUSINESS RISKS TO THE CAPITAL FINANCING PROCESS

Water utilities face significant requirements to comply with increasing regulations, replace aging infrastructure, fund employee health benefits and pension obligations, and meet growth-related increases in demand. Risks associated with meeting these requirements need to be managed to reduce the variance between anticipated outcomes and actual results.

A broad definition of risk is the possibility of something unexpected occurring and may result in a positive or negative outcome. However, the following discussion will deal primarily with the risk of negative outcomes that management should be aware of. Accordingly, risk may be further defined as:

- The possibility of suffering harm (a danger)
- A factor, thing, element, or course involving uncertain danger (a hazard)
- The possibility or an unacceptable probability of loss or harm

Risks associated with asset management were discussed previously; however, to achieve capital spending objectives, management will also need to consider other business risks to the utility. Business risks may be defined as "the threat that an event or action will adversely affect an organization's ability to achieve its business objectives and execute its strategies successfully." Business risks are more general in nature and come primarily from external factors that utilities face such as financial, political, stakeholder, environmental, security, and regulatory risks. Failure to consider these more general risks could result in adverse impacts on customers and/or be detrimental to the utility's mission and goals. Business risks facing water utilities include the following:

- Regulatory mandates
- Construction
- System size and customer base
- Local economy and customer base
- Governance and management quality
- Strategic focus
- System source water and capacity
- Asset maintenance
- Regulatory compliance
- Rates, rate structure, and rate-making flexibility
- Liquidity

Dealing with business risk requires answers to three questions:

1. What major risks will the utility face?
2. To what extent will the utility be impacted positively or negatively with each identified risk?
3. What is the probability that gain or loss will occur with an identified risk?

Risk measurement plays a central part in answering questions 2 and 3 because only after uncertainty is quantified (i.e., when a probability is assigned to an uncertainty) can risk discussions become useful. Utilities may want to combine traditional risk analysis with other innovative methods such as Monte Carlo, Consequence or Probability modeling; Failure Mode Effect, What If, Fault Tree, or Event Tree analysis; and others to calculate the degree of risk.

Regulatory mandates. Can the water system meet current and future regulatory mandates? Regulation is frequently the driving factor for capital improvements. The primary tool for addressing this risk is a long-range infrastructure master plan that identifies necessary capital improvements to meet current national mandates and water quality requirements as well as future goals that are more stringent than current regulatory standards.

Construction risk. Can the project be constructed on time and on budget with full capabilities? For success, the key tools are using proven designs, realistic schedules, managed startup risks, and reasonable cost estimates. Risk can also be reduced when new construction and infrastructure replacement are coordinated with other public improvements, such as road and sewer construction.

System size and customer base risk. Can the system and customer base support planned capital improvements while also fully funding O&M? A water utility is limited in its opportunity to change the customer base. However, system size risk can be reduced by ensuring that improvements to support growth do not burden existing customers; seeking the benefits of a larger utility by partnering with other utilities; making process improvements that reduce O&M costs and free up revenues to fund necessary capital improvements; extending capital improvements over several years to minimize rate impacts; and ensuring that any additional systems acquired by purchase or merger do not burden existing customers, but pay for themselves.

Local economy and customer base risk. Can the water system meet the changing needs of the local economy and population? Limited options are available to a water utility regarding the composition of its customer base. This risk can be reduced by ensuring that business development initiatives are cost-based and provide, where possible, diversity in the customer base to reduce dependency on any one sector of the local economy or customer class.

Governance and management quality risk. Are management practices institutionalized, recognized by political leaders, and able to withstand key personnel changes? Of all of the risk factors identified, this has the most opportunity for management influence as well as greater risk from inattention. As the workforce ages and retires, utilities will experience more frequent turnover in key management and staff. As this occurs, can a utility continue to develop long-term strategic and operational plans, participate in the AWWA Utility Benchmarking Program, involve customer feedback in key decisions, ensure professional certifications for staff, and withstand routine audits of financial results? Well-documented transition plans and staffing overlap are two ways to reduce this risk.

Strategic focus. Does the water system adequately anticipate asset maintenance, upgrades, and expansion? This requires assessment of the condition of the water system, clearly identifying system performance objectives, with strategies and funding to accomplish the objectives.

System source water and capacity. Are there possible threats to the water source from extreme weather patterns, fires, earthquakes, volcanic activity, aquifer contamination, or other jurisdictional encroachments? Will the water system be able to meet current and future maximum-day and maximum-hour demands? While natural disasters and threats to source water are difficult to plan for, mitigating alternatives will still need to be considered. Long-range facility plans that identify these mitigating alternatives and the related capital requirements are essential for system construction, as well as financial and rate planning. Consistent measurement and monitoring of demand patterns for hydraulic modeling and system planning can also support cost-of-service ratemaking.

Asset maintenance. Is the water system maintained to operate in perpetuity? This may be the single largest risk issue facing older water systems. Water systems were designed and built to a wide spectrum of useful lives using existing materials and design standards. The objective is to provide the continuous investment necessary to operate the system in perpetuity. This requires an inventory of aboveground and buried infrastructure, including condition and performance history. Preventive maintenance programs can keep assets operating successfully over their design life, but infrastructure rehabilitation and replacement will also be needed to ensure continued service into the future. Implicit in this is a determination that there should be no long-term deferred maintenance.

Regulatory compliance. What is the regulatory compliance record of the water utility? What is the impact of funding employee health care and pensions on utility operations? A time-consuming but valuable task is to compile an inventory of regulations affecting your utility and assign responsibility for compliance know-how to at least one specific individual for each. In addition, it is critical to rapidly and accurately respond to each occasion of a notice-of-violation.

Rates, rate structures, and rate-making flexibility. Do rates and charges reflect all financial commitments of the water system? The primary tools for managing this risk are regular cost-of-service rate studies based on best practices, long-term financial planning that includes all financial requirements, and full-cost pricing with no interclass rate subsidies.

Liquidity. Are revenues adequate to cover debt service and both recurring and one-time charges? This risk can be managed through financial modeling to forecast cash balances, complying with minimum debt-service coverage and investment requirements, timing of investment maturities to meet capital requirements, and insurance programs that provide a full range of liability coverage.

Managing and reducing risk. In addition to these practices, water utilities can incorporate a combination of preventive controls to keep risk events from occurring, detective controls that alert management to problems and irregularities that have already occurred, and corrective actions for those errors that are detected. Additional guidance can be found in chapter 20 of AWWA's *Financial Management for Utilities* and the 2013 *Internal Control—Integrated Framework*, by the Committee of Sponsoring Organizations of the Treadway Commission.

In a changing and unpredictable world, risk management provides a strategically based, comprehensive framework for identifying risks, avoiding pitfalls, and seizing opportunities to grow stakeholder value. How utilities perceive the probability of something occurring may determine how they deal with the event should it occur. When equipped with an awareness of potential events, management can foresee the likely effects of these events on future performance. By developing a better understanding of risks in relation to the strategies pursued, resources can also be used more effectively, and confidence in the utility's ability to perform as expected is increased.

Risk management also enables a utility to communicate more effectively with the investment community about the risks taken and the potential variations in the utility's performance, thereby reducing uncertainty and lowering the cost of capital. Utilities that understand risk and manage it strategically have the best chance of striking the optimum balance between risk and reward. Doing so is fundamental to creating value.

SUMMARY

This chapter describes the overall process of capital financial planning and sets the framework for more in-depth discussions of each of the steps presented in subsequent chapters of this manual.

Capital financial planning is a highly integrated and iterative process that begins with a strategic plan. Developing a strategic plan requires an evaluation of internal and external factors and the setting of prioritized and measurable goals. One important internal factor is the condition of utility plant assets and related risk to the utility of their failure. This leads to integration with a risk-based asset management program and an infrastructure master plan. Two very important external factors are future water supply and demand and new regulations. Goals should be prioritized and provide answers to the questions: "Are we doing what was planned?" and "Are we achieving the intended results?"

This leads to the development of multiyear capital improvement and related multiyear financial plans. The multiyear CIP shows how the utility will achieve its infrastructure master plan and asset management goals over the next three to five years. The multiyear financial plan shows how the utility will finance continued operations as well as planned capital expenditures over the next three to five years. Both of these plans work together to provide a road map of the utility's intended capital expenditures and related financing sources—including impacts on customers and stakeholders. After adjusting for the impacts of business risks, these plans are implemented through the utility's annual budget process.

Chapter 2 goes into more detail on identifying various financing alternatives and chapter 3 discusses factors to consider when evaluating each alternative.

BIBLIOGRAPHY

American Water Works Association (AWWA). 2012. *Financial Management for Water Utilities: Principles of Finance, Accounting, and Management Control.* Ann T. Bui, Editor. Denver, Colo.: American Water Works Association.

AWWA. 2012. Manual of Water Supply Practices M1, *Principles of Water Rates, Fees, and Charges,* Sixth Edition. Denver, Colo.: American Water Works Association. www.awwa.org.

AWWA. 2007. Manual of Water Supply Practices M50, *Water Resources Planning,* Second Edition. Denver, Colo.: American Water Works Association. www.awwa.org.

Committee of Sponsoring Organizations (COSO) of the Treadway Commission. 2013. *Internal Control—Integrated Framework.* Altamonte Springs, Fla.: COSO. www.coso.org.

Goetz, Melanie K. 2014. *Communicating Water's Value: Tips, Talking Points, & Strategies.* Denver, Colo.: American Water Works Association. www.awwa.org.

Chapter 2

Identifying Financing Alternatives

Each time a utility's financial plan is updated, utility management should review its financing alternatives for capital projects. Management's goal should be to identify the alternative or combination of alternatives that provides the utility with funds needed to support the necessary capital program at the most acceptable cost recovery. The question to answer is: How can capital improvements be paid for without causing rate shock to the utility's customers?

This chapter examines various types of capital funding and financing alternatives available to water utilities in the United States. Capital improvement financing alternatives typically include internal sources, such as cash from operating revenues or accumulated reserves investment, as well as external sources, such as debt, grants, and leases, or a combination thereof. Privatizing and outsourcing of some operations may also be acceptable substitutes for some types of capital investment. After specific financing alternatives have been identified, the utility should explicitly evaluate the advantages and disadvantages of each source of funds. This evaluation process is discussed in more detail in chapter 3. Because financial markets continually change, this list of alternatives is not meant to be all-inclusive.

INTERNAL FUNDING

In the United States, many utilities pay for some system improvements with internal funding. Because of the implications to rate stability, this "pay-as-you-go" funding is often used to pay for smaller or short-term capital projects, or to reduce new debt interest expense related to large capital projects by partially funding construction costs from revenue. Internal sources of cash can include dedicated tax revenue, water rate revenue, impact fees, system development charges (SDCs), and contributions from developers and customers. More extensive information on SDCs can be found in AWWA Manual M1, *Principles of Water Rates, Fees, and Charges*.

To maximize the portion of operating revenues available for capital financing, utilities often look at one or more of the following:

- Cost reductions and cost avoidance
  - Process improvements
  - Energy cost savings
  - Outsourcing activities
  - Value engineering reviews of capital projects
- Revenue enhancements
- Depreciation
- Inter-enterprise or municipal agency loans

Cost Reductions and Cost Avoidance

Reviewing current business and operational practices can uncover ways to reduce the cost of those practices or ways of avoiding expenses altogether by increasing efficiencies.

Process improvements. An operational review to maximize the efficiency of an organization can often free up operating funds to be used for capital improvements. Process improvement involves identifying key business processes associated with operations and evaluating the effectiveness of each process. Once each process has been reviewed, operations can often be streamlined by eliminating wasteful practices, introducing standard practices, providing better training, or increasing use of automation. Such process improvements could reduce utility costs by decreasing nonrevenue water or enhancing and improving proactive maintenance programs. Opportunities for savings, such as through competitive bidding or the use of fixed-cost annual contracts, are often found as a result of this business process review. The adoption of a formal asset management system can also reduce capital expenditures over time by focusing on maintenance, repair, and replacement activities and their associated budgets to produce the lowest life-cycle costs for major assets.

A high number of approvals required in the course of a process often indicates inefficient business processes and that an operations review is needed. Additional approvals may have been incorporated into a process that did not consistently produce the desired results to catch costly errors. As processes evolve, inefficiencies such as duplicate tasks, incremental work steps, or outdated methods often become ingrained, adding time to the processes. This additional time usually results in increased labor or duplicated costs. By eliminating unnecessary steps and improving the remaining steps, a utility can improve the process itself, thereby spending less money on materials, equipment, and labor for associated activities. Such savings will release funds that can be used to pay for capital improvements. Process improvement effectively reduces costs by evaluating these tasks that may have become burdened by inefficiencies over time.

Energy cost savings. Energy to operate treatment and distribution-system pumping facilities represents a significant operating expense for water and wastewater utilities. Water utilities are typically among the largest customers of a local energy provider. Rate structures employed by electric utilities often include substantial demand charges for high energy use during peak usage periods. A demand charge poses a problem for utilities because customer demands for water service and electric service often coincide. Such demands lead to the very result that the utility wants to avoid: high electric usage during peak periods.

Solutions for high electric demand charges include better management of system operations through the use of supervisory control and data acquisition (SCADA) systems, written policies, and operator training. Petroleum or propane gas-driven pumps can be installed and used during peak electric use periods to reduce demand charges or to prevent the utility from being placed in a higher rate bracket. Distribution system storage tanks can be filled during off-peak hours so that additional pumps do not have to be turned on during peak hours to address this need. As energy costs climb, the implicit ratios and trade-offs inherent in the design of capital facilities need to be examined and perhaps updated.

Electric and gas companies generally employ multiple rate structures. In most cases, a utility could potentially meet the criteria for several structures. By becoming familiar with the structures and negotiating for the most advantageous one, a water utility can realize significant savings.

In addition, some energy utilities have different rate incentives for different types of fuel. The water utility should account for these rate differentials when analyzing what types of pumps to use. Converting from one fuel source to another may be economical because the energy utility may be offering incentives that cover some or all of the conversion costs.

Recent developments and technological advancements in the field of Smart Energy may offer benefits to a water utility. Smart Energy refers to technologies that support the ability to send remote pricing signals and varying tiered-rate structures to customers. Automated metering infrastructure allows two-way communication between the user and the provider that can support moving energy usage from peak usage periods to non-peak periods.

With the deregulation of electric and gas rates, it may be advantageous for a water utility to examine the feasibility of joining a purchasing consortium or pool, to negotiate long-term power or gas rates as part of a larger group of energy consumers.

Outsourcing activities. Water utility operations include a wide variety of activities. While many of these activities may be performed more cost effectively by the utility, some activities may be more efficiently and effectively performed by others. Such activities should be considered for outsourcing. An example would be when significant capital outlays are required for equipment that is only seldom needed. Private businesses may be able to buy this equipment and contract with several entities to use it more fully. Because the cost of the equipment is spread over several entities, everyone can save money. Sharing of seldom-used equipment or facilities by two or more water utilities could also reduce costs.

Another example is contracting labor to handle peak activity that base staffing levels cannot support. An engineering division can be staffed at a level to handle normal design activity, and consultants can be employed when design and construction levels exceed the capabilities of the base staff.

A useful exercise when evaluating a utility's efficiency in performing an activity is to determine the cost of outsourcing that activity and then require the utility to meet or beat that cost. If the utility cannot operate at a cost that is competitive with outside entities, the utility should either reengineer the process to become competitive or contract the function. Some utilities have internal groups formally bid against outside contractors as a way to maintain cost-sensitive pressure on the utility.

Value engineering of capital projects. If a water utility has large dollar capital projects in its capital improvement plan (CIP), management may benefit by engaging an engineering firm, other than the design firm, and a general contractor to independently review the design of the capital project before putting it out to bid. This process is called *value engineering* and will result in suggestions to improve efficiencies and reduce associated costs of the project. The use of value engineering is a best management practice for large construction projects, and spending a relatively small amount for a second opinion can often produce significant savings in construction costs. Additionally, if value engineering determines that project implementation is not an immediate need and may be postponed, a significantly smaller impact on utility finances may be achieved.

Revenue Enhancements

Most utilities in the United States receive general revenue from several different stakeholders and investments. Stakeholders include

- developers (to support system expansion),
- builders (to offset the cost of installing new meters or distribution lines and facilities), and
- water customers (to pay for the cost of treating and delivering water).

Utilities should establish fees and charges that recover the cost of providing service to each group to prevent unintentional subsidizing of one group by another. Also, efforts should be made to ensure the utility's billing function captures all service provided. For example, the utility's engineering support provided to developers and builders for reviewing utility aspects of their site plans and for installing service connections and meters should be billed based on site needs. Programs should be established that minimize nonrevenue water, beginning with a water audit that can provide a reliable assessment of water loss. The water audit process is described in AWWA Manual M36, *Water Audits and Loss Control Programs.* Such water is costly because money is spent to obtain and treat water that is never sold. Revenue can be increased or costs reduced through testing and changing out faulty meters or through a leak-detection survey to identify needed repairs. More extensive information on water meters can be found in AWWA Manual M6, *Water Meters—Selection, Installation, Testing, and Maintenance.*

Most utilities receive interest earnings on invested idle funds, either through daily investment of funds needed in the short-term or longer-term investment of funds needed in the future. Interest rates vary by length of investment. Careful management of idle funds investment, especially longer-term investment of funds intended for capital projects, should provide additional revenue for the utility that minimizes future revenue needs. A municipality can earn higher rates of interest on longer-term investments of its revenue sources. For example, customer contributions in the form of system development charges may be invested at a higher yield than daily pooled investments until they are used for capital projects.

Federal law regulates the rate of interest earned on borrowed funds. If a municipality must have funds on hand before authorizing project construction, a large project requiring more than one year's building time can generate significant interest earnings. Such interest earnings must be recognized in the sizing of the associated debt issue. A smaller debt issue may reduce future annual debt service requirements.

Depreciation

Annual depreciation expenses for existing capital is an accrual-based cost that may be appropriately recovered through water rates. This can be accomplished either by reflecting the actual accrual expense or as a systematic return of owner's invested capital through water rates. While depreciation is a noncash expense and is typically not included in the revenue requirement under the cash needs approach, the expense is often used to roughly approximate the minimum amount of needed repair and replacement investment to existing assets that may be necessary to maintain the condition of the system. If appropriate levels of annual depreciation are used to set rates and those internally generated funds are set aside for the replacement of worn-out capital infrastructure, a utility will be in a position to maintain its financial integrity and limit the use of debt financing.

If the equity or utility basis method of determining revenue requirements is used, then depreciation is included as part of the revenue requirement in lieu of debt service and pay-as-you-go capital expenditures, which costs cannot be included as cost-of-service to prevent the double recovery of renewal-related capital costs (see Glossary for cost-of-service). Chapter 7 describes the equity method of determining revenue requirements. If depreciation is used as a cost element, the return-on-rate base should also be included as a portion of the revenue requirement to provide compensation for borrowing costs and risk. Failure to include return as a cost-of-service component could cause serious cash flow problems for the utility because the combination of depreciation and return is intended to cover a utility's annual capital funding needs.

Investor-owned utilities and government-owned utilities whose rates are regulated by individual states are generally required to employ the utility basis cost-allocation methodology in their rate setting. However, state regulatory agencies limit the amount of depreciation and the investment on which a return may be earned to the level of original cost assets that the owners have built with the owners' resources. Although investor-owned utilities own the assets under their control regardless of funding source, regulatory agencies do not want investor-owned utilities to profit from the investment in fixed assets contributed by others. When investor-owned utilities replace contributed assets with facilities funded by the owners' resources, the utilities may recover depreciation and earn a return on such facilities.

Government-owned utilities in the United States are generally self-regulated and may define depreciation and return as applying to all of their assets, regardless of funding source. When such utilities accept contributed assets, they agree to operate, maintain, and replace such facilities. However, water rates are the major source of revenue for government-owned utilities. By including all annual depreciation and earning a return on a rate base that includes all undepreciated portions of assets in its rate calculations, government-owned utilities may generate sufficient funds to pay debt service on outstanding bonds and meet their recurring annual capital needs. The depreciation period is based on the useful life of the asset. If an asset is debt-financed, it may have a useful life (basis of depreciation) of 50 years, but a bond term of only 20 years. Under this circumstance, depreciation for a specific new asset may not produce sufficient cash to repay, in a timely manner, the principal portion of the debt on that asset. However, annual depreciation is generated from all assets, many of which may not have debt associated with them. Unless a utility is highly leveraged, total annual depreciation should be sufficient to cover the total annual principal costs of outstanding debt.

Municipal Enterprise Fund or Agency Loans

Frequently, municipalities operate multiple utilities, such as water, wastewater, and stormwater utilities. A more financially stable utility, which has temporarily available surplus operating or capital funds, could make a short-term, 2- to 5-year loan to a utility attempting to become revenue self-sufficient and at interest rates similar to the investment earnings rates that the municipality receives on its investment of idle funds. The loan may help the receiving utility to phase-in new rates over time as it improves its financial position. If another municipal agency has available surplus funds, a similar internal loan could be made.

EXTERNAL FUNDING

As discussed in the previous section, internal funding mechanisms are frequently employed to finance some of a utility's capital needs and are an appropriate method for short-duration, low-cost routine capital projects. Utilities often pay for such projects on a pay-as-you-go basis (e.g., from existing customers through rates for service or from interest earnings on investment of utility funds). For more expensive major capital projects, however, utilities usually need to find an outside funding source. External sources can include the following:

- Debt financing
  - Common short-term debt
  - Common long-term debt
- Leasing
- Contributions
- Grants
- Tax benefit districts

To stay financially healthy, a utility should balance the manner in which it finances capital projects with a combination of internal and external funding. This section describes each of these external funding sources and their characteristics. Chapter 3 discusses some of the advantages and disadvantages of each and provides suggestions for evaluating each source of funding.

Debt Financing

Major capital projects are frequently funded with some form of debt. Debt financing allows a utility to raise more funds for capital projects than would be feasible on a pay-as-you-go basis, and it allows the utility to repay those funds over time. Debt can be repayable over various periods and spreads project costs to future, rather than current, users of the system. The length of time between borrowing and fully repaying determines whether the debt is considered long-term or short-term debt. Debt that must be repaid in one year or less is usually considered short-term debt. Debt that is repayable over longer periods may be called intermediate- or long-term debt.

The nature of the promise to repay debt may differ in various ways. Some debt may require that a portion be repaid each period; other debt may have periodic interest payments but require the full principal amount to be repaid at the end of the term. The utility may make a general promise to repay debt, or it may provide specific collateral as security for its repayment obligation.

Debt may be raised from various sources, both public and private. How easily a utility can access these sources, and how much it will be able to borrow, will depend on its size, its current financial condition and ability to repay the debt, and the impact the debt will have on its rates.

Common Short-Term Debt

Short-term debt may take the form of bank loans, notes, commercial paper (CP), or floating-rate demand notes.

Bank loans. Loans from banks can be structured as term loans or as lines of credit. A *term loan* is repayable at a specific future date. A *line of credit* is an arrangement under which the borrower can borrow and repay money from time to time as needed. A line of credit may be *committed* or *uncommitted*. According to an uncommitted line, loans are made at the bank's discretion with terms set at the time of borrowing. In contrast, with a committed line of credit, the utility and the bank enter into an agreement that (1) sets the terms under which loans may be made and (2) obligates the bank to honor requests for loans during the duration of the agreement, as long as the borrower meets predetermined credit criteria. A committed line of credit generally requires that the borrower pay a fee to the bank in return for having a source of funds available.

Bond-, tax-, grant-, and revenue-anticipation notes. A note is an unsecured debt, usually with a maturity under 10 years. The various types of anticipation notes are promises by a municipal issuer to repay the borrowed amount from a source of money that the utility will receive in the future, such as a future bond issue (in which case the note is called a bond-anticipation note or BAN), anticipated revenues (revenue-anticipation note or RAN), taxes (tax-anticipation note or TAN), or grants (grant-anticipation note or GAN).

Commercial paper and tax-exempt commercial paper. CP is an unsecured short-term promissory note with an average maturity of 30 to 45 days. The notes, which according to security regulations have a maximum maturity of 270 days, are intended to be refinanced (rolled over) continuously for periods that may exceed one year. CP may be issued by investor-owned or government-owned utilities. When CP is issued by a state or local government, it is called *tax-exempt commercial paper* (TECP). CP and TECP differ from other interim financing in that they are generally rated by a national rating agency and supported by a backup source of funds, such as a bank line of credit, a letter of credit (LOC), or another credit facility, to protect the purchaser of the notes from default in the event the utility cannot sell a new issue when a maturing issue becomes due. The use of TECP by municipal issuers has increased because of financial market demands. TECP is a prime investment instrument for tax-exempt money market funds.

Floating-rate demand notes. Floating-rate demand notes are securities on which the interest rate changes at predetermined intervals (often monthly) and that give the purchaser of the security the right to demand that the seller (utility) redeem the notes at each interest adjustment date. Floating-rate notes are usually issued in $100,000 denominations and are placed primarily with money market funds or other large institutional buyers.

Common Long-Term Debt

Long-term debt may take the form of a note, a debenture, or a bond. Both notes and debentures are unsecured borrowings. Notes generally have maturities under 10 years and debentures have maturities of 10 years or more. While the term *bond* in strict usage means a secured debt, in general usage, it can refer to various kinds of secured or unsecured debt. In this manual, bond refers to all types of long-term debt.

General obligation (GO) bonds. General obligation bonds, referred to as GO bonds, are debt obligations issued by a government entity, such as a state or local government, and backed by a pledge of the entity's full faith and credit. GO bonds can be issued only by units of government that have the authority to tax. Water utilities that are separate from local governments may not have this authority. GO bonds are secured by an unconditional pledge that the issuing government will, if necessary, levy unlimited taxes to repay the debt. This is the strongest pledge the government can provide, and it is usually regarded by investors and rating agencies as the strongest form of bond security. Therefore, a GO bond generally has the lowest cost of the various financing vehicles available to local governments.

Because GO bonds are issued by a state or local government, interest paid to the lender is usually exempt from federal income taxes. Depending on state tax laws, this debt may or may not be exempt from state income taxes. Municipalities should conduct further analysis if debt is used for any activity that may be associated with private use. Examples to scrutinize carefully are construction projects related to private–public partnerships and design–build–operate contracts. Utilities must ensure that constructed facilities are used for the public good to ensure tax-exempt status. Another important consideration between the use of GO and revenue bonds, which are described in the next paragraph, is that GO bond issuance often requires a public vote. Revenue bonds typically do not. At times, the public may vote down bond referendums. Utilities have the advantage in that they can usually issue revenue bonds without referendums.

Revenue bonds. Revenue bonds are debt obligations issued by local governments or other public agencies. Principal and interest on revenue bonds are secured by (and may be payable only from) the specific revenues named in the bond documents. Revenue bonds are commonly used to finance activities that generate revenue and that are expected to be self-supporting, such as a water utility. Periodic debt-service payments on revenue bonds issued by water utilities are usually paid from water rates and miscellaneous revenue, although all system revenues typically secure the bonds by covenant.

With a revenue bond, debt repayment may depend on timely completion of a project, on adequate rate or charge structure, and on sound fiscal management of the enterprise. For this reason, the bonds are assumed to be riskier and will have a higher interest rate than a GO bond issued by a comparable issuer. Investors' acceptance of revenue bonds is highly dependent on the service or project to be financed and on the creditworthiness of the borrower. The spread (difference) of interest rates between GO bonds and revenue bonds has historically varied. More recently, municipal bankruptcy has cast doubt on the strength of the "full faith and credit pledge" versus the security interest that is given in a revenue pledge, and in general, investors have favored water revenue bonds because of their essential nature and see them as less vulnerable to cyclical downturns associated with general governments.

Most revenue bond agreements or indentures contain provisions, or covenants, that are intended to enhance marketability by providing investors with additional assurances that their money will be repaid. The following are some provisions typically included in revenue bond documents:

- A rate covenant, by which the issuer agrees to set rates sufficient to meet all operating costs and some multiple of debt service. For example, a requirement may be that net revenues available after operating costs will be sufficient to cover 125 percent of debt service. The extra 25 percent may be available to contribute to reserves or for funding other capital costs needed to keep the system in good financial and working order.

- A flow-of-funds requirement that specifies the order in which revenues will be used. For example, there may be a requirement that revenues be used first to pay O&M expenses, then for debt service, then to replenish reserve funds, and then for costs associated with system repairs or replacement. The required reserve amount may decrease over time, freeing funds for other capital needs.
- The establishment of a fund to set aside money to ensure debt service payments are made on a timely basis or to cover debt service for some period of time in the event of a revenue shortage. This deposit of money is called a *debt service reserve fund,* and the indenture generally requires that the fund be administered by a third party, such as the trust department of a bank. For example, many agreements require the reserve amount to be equal to the maximum debt service in any year during which the bonds are outstanding.
- A parity test, which provides that additional bonds may not be issued unless historical and projected revenues indicate that there is sufficient revenue to avoid dilution of coverage on outstanding bonds. The test is typically testified as to its sufficiency by the chief financial officer or by an independent third party, such as an auditor, feasibility consultant, or consulting engineer.

Sometimes a government adds a GO pledge to a revenue bond backed by a specific source of revenues, such as the net revenues of a water system. These bonds are referred to as *double-barreled bonds* because they provide the dual security of a dedicated revenue source with the general taxing power of the government.

Bonds with credit enhancements. A municipality or utility may have difficulty issuing debt, especially if it has not issued debt before, does not have a credit rating, and is unknown to the investment community. One option available to this type of organization is to borrow from a local bank or other institution where the organization's creditworthiness is more easily demonstrated.

Another option is to obtain a credit enhancement, generally in the form of bond insurance or an LOC from a well-respected and financially sound entity. Under such an arrangement, the utility will pay a fee to have the financially sound party, such as an insurance company or bank, provide a guarantee to the investor in the bond's official statement that it will ensure the timely payment of principal, interest or both. The addition of a guarantee by a large, well-known entity with high credit ratings makes a bond more attractive to investors, resulting in a lower bond interest rate.

Strong municipal financial and economic positions and high bond interest rate levels are also situations where credit enhancements may warrant consideration. Credit enhancements may add a level of stability to the entire municipality's financial situation. Some municipalities may employ credit enhancements regularly as a matter of policy. When interest rates are high, credit enhancements, depending on the creditworthiness of the issuing company, may result in the highest credit rating for the municipality and a lower bond interest rate.

Unless the debt interest cost over the life of the bond, with a credit enhancement, plus the cost of the credit enhancement is significantly less than total debt interest costs without enhancement, third-party credit enhancements may not be appropriate. Utilities should evaluate the savings in debt interest cost over the life of the bond, with and without credit enhancements. If there are no savings, the utility may want to consider other factors in making its decision, such as the municipality's historical financing practice or policies. Employing credit enhancements is as much of a policy issue as it is a cost issue.

Interim (temporary) financing. It is common for a utility to begin construction of a capital project before the final construction cost for the project is known. In such cases, interim (temporary) financing is often used. Typically, a lender will issue a line of credit to be used to pay for the construction costs, and the utility will draw on the line of credit as needed. When construction is finished or when the utility has a reliable estimate of the final construction cost for the project, the interim financing is paid off by including the amount borrowed on an interim basis in the total amount of the long-term debt.

An alternative would be to issue consecutive bonds with each covering a portion of project construction. The first issue covers a significant portion of estimated costs that permits the initial phases of construction. The second issue, called a *completion bond,* covers all remaining costs of the project when they are better known. Purchasers of the first issue are notified, in the bond's first official statement, that a completion bond will be issued. The staging of bonds has an additional benefit of permitting the utility to phase-in rates over a period of two or more years to cover both issues.

Governmental loans. Federal and state agencies provide a valuable source of lower-cost financing through a number of programs that supply infrastructure financing. The federal Rural Utilities Service (RUS, formerly the Farmers Home Administration) is the most common provider of loans to rural and economically depressed areas for water and wastewater system improvements. Amendments in 1996 to the Safe Drinking Water Act establish a state revolving fund (SRF) program to provide loans to water utilities for capital improvements at interest rates at, or less than, market rates. A similar program under the Clean Water Act has supplied low-rate loans for wastewater treatment facilities for many years. On June 10, 2014, the Water Resources Reform and Development Act (WRRDA) was signed into law. This Act provides authorization for the Water Infrastructure Finance and Innovation Authority (WIFIA) five-year pilot program. This program will provide loans for large ($20 million or more) water, wastewater, and stormwater infrastructure projects at US Treasury rates. The utility's financial advisor should be aware of current sources of governmental loans and should be consulted about the structure and terms of loans available to the entity.

These programs allow utilities to avoid many of the steps associated with an open market debt offering and are prevalent in many states, especially those with smaller or less tax-advantaged municipal debt markets. While these programs are popular, especially for smaller utilities, they often are accompanied with inflexible amortization schedules, significant fees from the sponsoring states, and may not offer the most cost-efficient funding available to the utility. In addition, each state fund will have legal requirements establishing the nature of the obligation.

Leasing

A *lease* is an arrangement between the owner of property (*lessor*) and the user of property (*lessee*) that gives the lessee the right to use the property for a defined period in return for meeting certain requirements, including the payment of rent. Leases may be attractive for municipal utilities as a means of acquiring the use of needed equipment and facilities when debt limitations restrict direct purchase and ownership. Leasing in some of its forms is treated as debt for financial reporting purposes, and leasing is generally viewed as a substitute for debt financing.

Because leasing often involves tax benefits to the lessor and tax implications to the lessee, and because it is subject to the interpretations of the US Internal Revenue Service, it is a complex form of debt financing. Tax acts passed in 1984 and 1986 substantially reduced the benefit of tax-oriented leasing. Utility managers should retain experienced legal assistance when considering a lease because tax laws are constantly changing.

Sources of lease financing include large banks, equipment manufacturers, real-estate development firms, and major leasing companies. Most firms provide preliminary lease-rate quotes, which may be used to compare lease costs against ownership costs. The major categories of leasing likely to be encountered by a finance staff that is considering how to finance a new facility or new equipment include direct leases, leveraged leases, and certificates of participation.

Direct leases. With a direct lease, the lessee negotiates the equipment or facility specifications, as well as the rental terms, with the lessor. After the lease is signed, the lessor purchases the equipment or constructs the facility exactly as specified in the lease agreement. When the equipment or completed facility is delivered, the lessee ensures all specifications have been met, then formally accepts the facility before making payments.

Leveraged leases. A leveraged lease is similar to a direct lease (which is sometimes called a nonleveraged lease), but it is more complex because another party is involved. The additional party is the entity that provides funds to the lessor for constructing the facility or acquiring the equipment.

Certificates of participation. A certificate of participation (COP) is a security backed by an interest in a stream of rental payments. The lessor is generally a governmental entity, often one specifically created to enter into lease transactions, and payments received by certificate holders are generally tax-exempt.

The financial advisor should be able to explain the current treatment of a COP transaction under any local debt statutes. The advisor should also be able to explain the current market requirements with respect to common restrictions concerning use of property, the flow of funds, security for the certificate holder, and the basis for tax exemption. Restrictions placed on the lessor concerning use of property and disposition of lease payments establish security for the certificate holder and the basis for tax-exempt interest payments from local government.

Contributions

Customer contributions of money or property are a common and important form of capital for water utilities. Historically, customer contributions have been related primarily to customer capacity in supply, treatment, storage, and distribution system facilities. Customers are frequently required to purchase their meters and pay for the installation of taps and services. Developers are often required to provide distribution mains, fire hydrants, and customer connections for new subdivisions as contributions in aid of construction. While these contributions relieve a water utility of some of its capital needs, they often do not provide financing for major supply, treatment, or distribution facilities.

Some utilities require developers to pay customer advances for construction of subdivision facilities. The utility may reimburse the developers during a specified period as new customers connect to the system. Any funds remaining at the end of the period are retained by the utility as contributions in aid of construction.

Government-owned utilities may also include funds provided through intergovernmental transfers as contributions if those funds are not intended to be repaid from utility revenues. This type of contribution is often used to cover a portion of capital requirements for specific facilities, and the funds may be provided by the local government from GO bond proceeds, property, or sales tax revenue.

Grants

Development of public water supplies in the United States has traditionally been a local responsibility. Unlike the wastewater industry, the drinking water industry has not had a large federally funded grant program. With limited federal spending and state revolving loan programs, a conservative financial plan does not count on federal grant money. This is true even as federally mandated drinking water quality standards require additional capital investment by local utilities.

However, a few grants from various federal and state agencies exist for physical plant security and rural or economically depressed areas. For example, RUS, operated under the auspices of the US Department of Agriculture, provides grants and loans for rural water and sewer systems and communities with populations of fewer than 10,000. The ratio of grant-to-loan funding from RUS is a function of the median household income of the population served.

Tax Benefit Districts

Many local governments have employed tax-benefit districts to cover the costs of local water distribution facilities serving specific portions of a utility's service area, and governments in several states employ them, including Alaska, Connecticut, Florida, New York, and Virginia. For example, by majority vote of the benefiting property owners, a local government could establish several tax benefit districts to initially fund, operate and maintain, and replace local facilities. The local government may also establish and own backbone facilities for supply, treatment, transmission, and storage to serve the benefit districts and charge customers within these districts on a wholesale water rate basis. In some situations, multiple tax-benefit districts could be provided with wholesale water service by a nearby incorporated community within a county. In other situations, large populated counties adjacent to large cities may fund all or a significant part of their water systems' capital and operating costs from a single tax benefit district.

When raising water rates to fund infrastructure replacements is met with customer and policymaker objections, employment of tax benefit districts to address critical local facility replacement needs in specific portions of a utility's service area may be a logical alternative.

Public–Private Partnerships

Public–private partnerships are contractual arrangements between public agencies and private companies. Public–private partnerships create benefits as they balance the resources, risks, and rewards of both the public agency and private company and provide greater efficiency, better access to capital, and improved compliance with a range of government regulations regarding the environment and workplace. An in-depth discussion of public–private partnerships may be found in chapter 3.

SUMMARY

In the United States, many sources of funds are available for capital improvement funding. Among these are internal sources, such as savings from improved operations and revenue enhancement, and external sources, such as grants, loans, and tax-benefit districts. Because of the size and long-term usefulness of capital assets associated with most CIPs, government-owned utilities generally rely on debt for a large proportion of their capital financing needs. Investor-owned utilities typically rely on reinvestment of utility net income, new investment from additional stockholders, and customer and developer contributions to meet their capital needs.

Chapter 3

Evaluating Financing Alternatives

The legal environment, financial considerations that affect all issuers, issuer considerations specific to the issuer, and other external factors all play a part in whether a particular funding alternative is the best choice for a utility. Legal factors may restrict the size and type of debt that a utility may assume. Financial factors, including the condition of the capital markets, could result in limitations on the debt size and interest rate. Issuer considerations include factors specific to the issuing utility, and external factors include whether the alternative is likely to be accepted by governing authorities and the public. This chapter discusses these factors and some advantages and disadvantages of the various financing alternatives discussed in chapter 2.

LEGAL FACTORS

One of the first steps in evaluating any financing alternative is to determine whether a particular option is permitted under the various laws, regulations, and governing documents to which the utility is subject. The utility's financial advisor and bond counsel, in conjunction with in-house counsel, will be able to discuss the various legal factors that are relevant to the availability of financing alternatives. If an alternative being considered is not currently permitted, the utility may want to evaluate other alternatives or the feasibility of changing its legal operating environment to include the alternative. Certain factors such as cost, effort, probability of success, and risks can be evaluated against the benefit that the utility would gain if the alternative were available.

An investor-owned utility may be subject to rules promulgated by the local regulatory body, as well as to previous indenture requirements, corporate resolutions, and restrictions in the utility's by-laws. A government-owned water utility is often subject to restrictions in its governing documents, as well as local and state/provincial laws. Both public and private utilities are subject to various federal laws if they wish to borrow on a tax-exempt basis.

Governing Documents

Legal factors affecting debt issuance are frequently found in a utility's own enabling documents, which typically include the original authorization and subsequent actions of the utility's board or other oversight agency. The governing documents for a government-owned utility often establish limits on borrowing, either as an absolute cap or as a ratio such as debt-service coverage. Governing documents may also identify what types of debt can be issued, including negotiated or competitive sales, restrictions on private placements, selection of the finance team, and other considerations. Most jurisdictions also have established rules about the investment of bond proceeds, with an emphasis on the preservation of capital assets.

Additional legal restrictions may be established by legislation or board action authorizing previous debt issuance. For example, a resolution may have already pledged certain assets as security for a previous debt issue. As a result, assets and revenues may not be available to support new debt unless the new debt is subordinated to the old. This may affect the feasibility and cost of the alternative.

State or Provincial Law

State and provincial laws in the United States and Canada frequently address the legal structure and borrowing capacity of local government entities, including municipal utilities and utility authorities. The requirements for voter approval of debt issuance, if any, are generally found in state and provincial statutes.

Some states require an additional level of review and approval before borrowing can be completed. One example of state involvement in municipal financing is the North Carolina Local Government Commission (LGC). A major function of the LGC is the approval, sale, and delivery of substantially all North Carolina local government bonds and notes. Before any unit of local government can incur bonded indebtedness, the proposed bond issue must be approved by the LGC. Some water utilities are regulated under their state's public service commission (PSC) or public utilities commission (PUC) and may require commission approval before borrowing. These regulations vary by states and whether the utility is private or public.

Federal Law*

The Securities and Exchange Commission (SEC) is the primary federal agency regulating utility financing in the United States. The SEC has extensive regulations that apply to public debt offerings made by investor-owned utilities. Through its regulation of underwriters, the SEC also mandates that public utility issuers make financial information available beginning at the time of borrowing and continuing until the debt is retired. The issuer's bond counsel often handles many of these activities; however, the issuer has the ultimate responsibility for ensuring compliance. In addition to financial reporting requirements, if a US based utility, whether investor-owned or government-owned, wishes to borrow on a tax-exempt basis, the utility will need to comply with arbitrage regulations of the US Internal Revenue Service, which address

* Please see chapter 8 for a discussion of Canadian utilities.

1. the maximum amount of interest that may be earned on borrowed funds after the transaction is completed and before the funds are expended on capital projects, and
2. how long the funds can be held before the issuer is subject to penalties and possible loss of the tax-exempt status of the issue.

Bond counsel and the financial advisor should be able to explain what is required from the utility so that it can assess the cost and administrative burden associated with federal requirements. It may be that the compliance requirements make the financing alternatives more attractive.

FINANCIAL FACTORS

Various financial factors can also affect a utility's financing decisions, including whether to pursue financing at all. While market conditions are beyond the control of the utility, these factors will play an important part in the decision-making process.

Financial factors that affect all borrowers include the current interest rate levels and terms, whether there is a general consensus about the level and direction of interest rates, the outlook for economic activity, and whether proposed legal or tax changes make borrowing more or less difficult than usual. Several of these factors are discussed in more detail in the following sections.

Market Conditions and Access

Utilities that plan to issue new bonds or refund existing bonds should be aware of the market factors that influence financial decisions. Bond markets have tremendous diversity. Corporate as well as municipal bond markets are influenced by the economic outlook, US Treasury activities, the shape of the yield curve* as presented in Figure 3-1, and general investor sentiment. In addition, municipal markets are influenced by proposed federal tax law changes, as well as by proposed changes in the requirements and specifications for issuance promulgated by the states or provinces.

If the markets are in an unsettled state, there could be advantages to using a negotiated sale or private placement (if available) rather than a public competitive sale because the issuer would have more flexibility in settling the time of sale. A *negotiated sale* is the sale of bonds directly to an underwriter and differs from a *competitive sale*, which requires

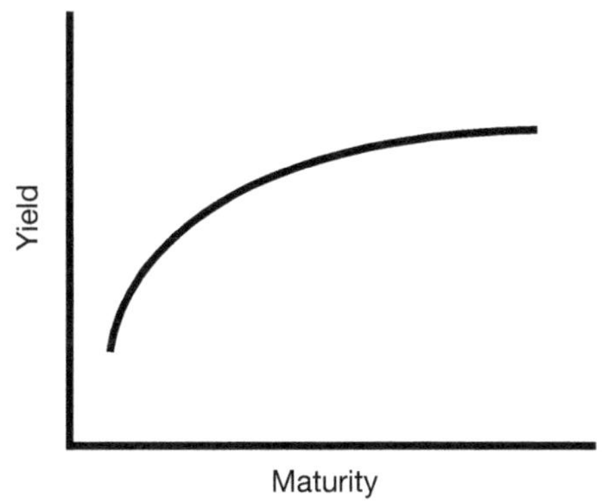

Figure 3-1 Treasury yield curve

* The yield curve shows the relationship between interest rate and time to maturity.

public bidding by one or more underwriters. Similarly, if the yield curve is flat, long-term interest rates may be comparable to short-term interest rates, which may influence the length of debt amortization. To properly assess the impacts of these and other market-related concerns and conditions, issuers typically rely on competent professionals in a variety of fields, including bond counsel, the financial advisor, and the underwriter. The assistance provided by these professionals is discussed in chapter 6.

ISSUER CONSIDERATIONS

Before making any decision regarding funding alternatives, a utility should carefully consider internal factors that influence which funding plan is most appropriate. If the plan includes new borrowing, financial considerations will be a factor in deciding which debt instrument is most appropriate. Among the financial factors specific to the issuer that need to be considered are

- whether the proposed alternative is compatible with the overall objective of the financing program, and
- whether selecting the proposed alternative will increase or reduce the utility's flexibility in the future.

As a part of its long-term financial plan the utility may have set targets for such things as the proportion of its debt that will be variable rate, the maximum amount of debt it will incur as a percentage of fixed assets, a maximum annual amount for debt service per connection, reserve funding requirements, or minimum debt-service coverage requirements in excess of indenture requirements.

Cash Versus Debt Funding

When considering funding options, one of the primary decisions that must be made is cash versus debt funding. The use of cash has benefits such as elimination of interest and issuance costs, reduced administrative effort, and decreased financial reporting requirements. However, to fully cash fund the capital needs of most utilities, rates would typically need to be set at levels that would be unacceptable to customers and governing officials. As such, the ability to assume debt is very important to a utility.

Other factors should also be considered when deciding between cash versus debt funding. Specifically, asset life should generally equal or exceed the term of the funding choice and recurring projects should be funded with recurring funds. Longer-life assets, such as treatment facilities and major transmission mains, are good candidates for longer-term financing. Shorter-life assets with recurring renewal and replacement needs, such as pumps, vehicles, and general equipment, are better candidates for cash funding. Many utilities establish rates such that levels of cash are generated each year to meet the utilities recurring renewal and replacement needs. Intergenerational equity is another consideration that should be given to the cash-versus-debt funding decisions. The use of debt financing can shift the burden of an asset's cost to future generations, thus requiring future customers to pay for projects completed in the past.

Meeting Covenants and Satisfying Revenue Requirements

The indenture requirements of existing debt are similar to the legal factors discussed previously. The indenture will contain provisions that affect a utility's ability to issue new debt. For example, existing bond covenants may specify that rates be established at a sufficient level to meet defined requirements, most often a debt-service coverage requirement. The indenture will provide details about bond covenants and the specific items to be included in the debt-service calculation.

In general terms, debt-service coverage is defined as annual net revenues (gross revenues minus operating and maintenance expenses) divided by total annual debt service (principal and interest on outstanding debt) with the result expressed as a percentage or times factor. For example, general obligation bonds may require debt-service coverage of 100 percent or a times factor of 1.0; senior lien revenue bonds may require debt-service coverage of 110 percent or 1.10 times; and junior lien revenue bonds may require debt-service coverage of 125 percent or 1.25 times. In addition to specific coverage requirements, rating agencies may expect the total of all utility debt service, including that for general obligation and revenue bonds and any other debt, to be covered by a minimum of 100 percent or 1.00 times.

A utility must be able to satisfy the covenant or be able to amend existing provisions when new revenue bonds are issued. When the amount of the new revenue bond debt is greater than the amount of all outstanding revenue bonds, the new covenants may replace existing covenants. In this case the new issue purchasers are assumed to have voted-in the new covenants with their bond purchases. Otherwise, an amendment may not be possible unless the outstanding debt is repaid or financially defeased (i.e., where sufficient front-end funds, including future investment earnings associated with the funds, are turned over to the control of a third-party paying agent to pay the debt service when it becomes due for the remaining life of the defeased debt).

However, if the plan is to repay outstanding debt, new investors may not be willing to accept a lower coverage ratio without a significant increase in the interest rate. A lower coverage ratio may result in the bonds receiving a lower credit rating. When evaluating what coverage is appropriate, investors consider customer base, local economy, historical financial performance, cash position, etc., as well as the demonstrated political will to raise rates when needed. Investors also consider historical data in conjunction with projections and assess the reasonableness of the assumptions on which projections are based. This evaluation is completed to determine the quality of the revenue stream pledged to pay bondholders. Investors and credit rating agencies will examine intra-period fluctuations and their causes, as well as the sensitivity of the revenue stream to adverse external events. With increased volatile revenue (which is an indication of increased risk), coverage requirements will often be higher to attain comparable investment protection comparable to issues with more stable revenues. The utility may want to perform a similar analysis

- to assess the reasonableness of being able to obtain a lower coverage ratio, and
- to answer questions that arise during the marketing process.

When developing its long-term financial plan and its ability to issue new debt, the utility should ensure that both its revenue and its cash needs are met over the planning horizon. The analysis should disclose whether sufficient working cash is available after the debt service on the new debt is considered. As noted in chapter 1, this is often an iterative process that balances financial requirements with financing options. For example, if utility rates are to be left unchanged for two or more years, the financial plan should show that sufficient reserves, such as rate stabilization and operating reserves, will be built up during the first half of the forecast period to finance an expected deficit in the latter half of the forecast period. If sufficient revenues are not shown, the planner may want to evaluate whether the utility's governing board would be willing to consider annual, biennial, or triennial rate increases, or restructure the new obligation in a way that fits better with existing obligations to enable the utility to meet both covenants and cash requirements during the forecast period.

Impact of Debt Portfolio

When considering forms of new debt, a utility should be aware of resulting changes to their debt portfolio. The issuer should start by understanding the amounts and types of debt currently outstanding. Understanding existing debt is critical to understanding how the new financing will influence, or be influenced by, existing obligations. Some specific concerns regarding the existing debt portfolio include:

- The structure of existing debt relative to new debt. How will the repayment terms of new debt fit with existing obligations? Are the same revenues pledged? Are payment and reporting dates the same? How will new debt impact administrative efforts? For variable-rate debt, when will rate changes occur on the various obligations, and what will this do to the organization's overall risk profile and revenue needs?
- The need and ability to issue subordinated debt (junior lien bonds) at different times for different purposes. Can new obligations be secured on par (equally) with existing senior lien obligations? If not, how will that affect the marketability of the new debt? What flexibility in bond covenants may be employed with the use of junior lien bonds? Is this flexibility available with senior lien bonds?
- The ability and desire to refund, or defease, existing debt in order to take advantage of improved interest rates and facilitate revisions to bond covenants. Are there restrictions in current indentures that could be revised with a refinancing? Have interest rates moved sufficiently so that refunding is economically justified? How will the limitations on advance refunding of tax-exempt debt affect what alternatives are available? Investor-owned utilities will also want to be aware of how the transaction will affect reported earnings per share and what effect this may have on a given utility's stock price.
- The impact of current indebtedness on the borrower's creditworthiness. Is the utility more or less leveraged than its peers? What affect will another debt issue have on credit ratings? How will a new debt issue affect the prices of currently outstanding debt? What affect will an adverse impact on outstanding debt prices have on the marketability of the utility's new debt (and, for investor-owned utilities, on stock price)? Does the utility have the ability to raise revenue, in the case of revenue bonds, and to meet the provisions of the rate covenant? What effect will the ability or inability to raise rates have on the utility's credit rating and ability issue future debt?

Rating agencies use financial ratios when evaluating a utility's creditworthiness. Utility management should ensure that financial ratios are developed for their financial plans and compared to other similarly sized utilities. AWWA's survey, *Benchmarking Performance Indicators for Water and Wastewater Utilities*, contains financial ratios that may be used to compare a system to other utilities. Chapter 4 provides additional considerations for preparing a utility for financing.

EXTERNAL FACTORS

Once the legal factors have been considered, and it has been determined that the utility is legally able to pursue the selected financing option and has the resources to repay the loan on a timely basis, certain external considerations must be addressed. Among these are whether the utility has public support for the capital improvements and for the proposed financing plan. These two matters are directly related, and the proposed financial plan often drives public debate on the need for the proposed capital program.

Most borrowings by government-owned utilities require citizen approval, either directly through referendum or indirectly through actions of an appointed board or elected council. Investor-owned utilities may need board approval, regulatory body approval, or both. Accordingly, the capital-financing vehicle should be selected with consideration for its acceptability to the general public, the utility's authorizing body, or both. Utility management should consider its ability to explain, justify, and successfully advocate the capital items to be financed, as well as the financial assumptions underlying the borrowing. A significant factor influencing public acceptance is the impact on customer rates. In jurisdictions where a ballot initiative is required for debt issuance, utilities should take particular care to evaluate the political climate and financial markets before approaching voters.

The need for public approval, whether directly or through elected or appointed representatives, should not deter a utility from using a financing mechanism that is appropriate and well-suited to a necessary and specific capital program. However, the likelihood of acceptance and approval of the proposed borrowing improves as financing alternatives are presented with greater clarity and understanding to public representatives. Interactive computer financial modeling tools can aid in the demonstration of short-term and long-term customer impact of alternatives and can help justify a financing recommendation. This review and reporting of operations, budgets, and forecasts provides an added level of security. However, borrowers should recognize that more complex and innovative financial products will face closer scrutiny.

The level of scrutiny for government-owned utility borrowing has increased since the financial market challenges of the mid-2000s. Rating agencies and borrowers have become much more critical in their review of a utility's financial projections, capital program, and management. Issuers who contemplate using derivative products and innovative or nontraditional financing should plan the proposed transaction to explain clearly to their boards or elected bodies. Even issuers who contemplate using traditional borrowing vehicles should thoroughly evaluate their financing decisions and be able to answer questions from overseers who are not familiar with such arrangements.

COMPARISON OF COMMON FUNDING ALTERNATIVES

This section describes some advantages and disadvantages of funding alternatives commonly used by water utilities, including general obligation (GO) bonds, revenue bonds, state revolving loan funds, leases, contributions, grants, tax benefit districts, bond structure variations, and public–private partnerships.

General Obligation Bonds

GO bonds are typically the lowest-cost form of debt financing available to local governments in the United States because these issues are backed by a pledge of the entity's full faith and credit. As such, GO bonds are generally a good source of financing for a start-up utility. Investors' familiarity with this traditional standardized form of financing makes GO bonds very marketable. However, more recently, municipal bankruptcy has cast doubt on the strength of the "full faith and credit pledge" versus the security interest that is given in a revenue pledge. Costs of issuance are generally lower than those associated with revenue bond financing, not only because the cost of marketing a bond with wide market acceptance is lower, but also because the documentation associated with a GO issuance is less complex than that required for revenue bond financing. Interest on GO bonds can also be capitalized during the construction period of the project, providing for lower or no debt service until project completion. Because GO bond financing does not require the restrictive covenants, special reserves, and higher debt-service coverage associated with revenue bonds, current revenues are not overly restricted or pledged, and can be used for other purposes.

While the issuance of GO bonds has many advantages, this form of financing also has certain drawbacks. One of the most significant disadvantages is that increases in the amount of GO debt must generally be authorized by the electorate. At a minimum, this increases the time needed to develop proceeds to finance capital projects. In addition, many states limit the amount of GO debt that can be issued or the interest rates that can be paid. These debt and interest-rate ceilings on GO bonds may be restrictive in areas where tax reform propositions have been passed. Additional state or local government restrictions may exist regarding the length of amortization of the bonds. This would tend to make the GO bonds' structure somewhat less flexible than other forms of borrowing. Finally, the utility must compete with other government agencies for available bond proceeds needed to fund their capital projects. Such competition may cause delays in the utility implementing its capital program.

Revenue Bonds

Revenue bonds offer important benefits to the issuer. They provide a mechanism for recovering costs of specific services, such as water supply, directly from the users that benefit from the services. Revenue bonds are available to agencies that do not have taxing authority to issue GO bonds. Unlike most tax-supported debt, revenue bond issuance does not typically require voter approval, although voters generally have a right to petition for a referendum. Revenue bonds also avoid possible dilution of the government's pledge of full faith and credit, because revenue bonds are not typically included in the issuer's debt limit. In addition, when issuing revenue bonds, the issuer has some flexibility to structure the new obligation in such a way that its annual debt service is structured around existing obligations. This form of structuring, known as *wrapping*, enables the utility to maximize its ability to meet annual cash requirements and debt covenants. Wrapping considers the structure of other existing debt obligations so the new obligation payments will have the least impact on the issuer. Similar to GO bonds, interest on revenue bonds can also be capitalized during the construction period of the project, providing for lower or no debt service until project completion.

Revenue bond financing also has some disadvantages. Because principal and interest on revenue bonds are secured by and payable exclusively from revenue received from system operation or the project being financed, market acceptance of these bonds is highly dependent on the systems' service capability, or project to be financed, and on the nature and reputation of the issuer. For this reason, revenue bonds generally command higher interest rates than do GO bonds. The additional cost of revenue over GO bond financing is subject to market variation and will also reflect the specific credit of the issuer. During the Great Recession of 2009 and 2010, the fiscal challenges facing general governments appeared to many investors to be greater than the enterprise risks faced by water and sewer utilities, and revenue bonds would at times be priced with lower interest rates than

GO bonds. To make an informed decision between the two, utility managers should seek assessments of the interest rate difference from market professionals.

Higher interest rates and marketing difficulties are particularly a problem for enterprises that do not have established earnings potential or a history of financial operations. Revenue bonds also frequently contain numerous provisions that potentially make revenue bond issuance more administratively cumbersome than issuing a GO bond. Indenture provisions, such as requirements for special reserve funds and debt-service coverage ratios and limitations regarding the amount of future debt that can be issued, may also decrease the issuer's future flexibility. These provisions often make it difficult to issue additional bonds and may trigger refinancing or refunding of outstanding debt earlier than the issuer had planned to retire the debt.

Finally, revenue bond borrowing is more complex and generally requires more time than GO bond borrowing. Revenue bonds usually have more complex legal, engineering, and trustee contributions. Therefore, the related issuance costs are also higher. Utilities financing projects with revenue bonds typically need to provide an independent demonstration of engineering and financial feasibility. The added cost of a revenue bond issue has important implications for an issuer. In some instances, the benefits of revenue bond financing may not be worth the added cost. A variety of factors, such as debt ceilings, issue size, rating, and flexibility of structuring, should be considered in making the choice between using revenue and GO bonds.

Government Loans

Government programs in the United States, such as the USEPA Drinking Water State Revolving Fund (DWSRF), Clean Water State Revolving Fund (CWSRF) programs, and the Water Infrastructure Finance and Innovation Authority (WIFIA), make low-cost loans and other types of assistance available to public water and wastewater systems. According to the SRF program, capitalization grants are awarded to states, which then utilize these funds to establish and administer state revolving loan fund programs. As payments of these loans are received, the principal is reissued in new loans, thus the name *revolving loan*.

Benefits of the SRF loans include below-market interest rates and potentially lower issuance costs. However, SRF programs are often very competitive, require compliance with state reporting requirements, cannot be creatively structured (such as including capitalized interest), and depend on the availability of funds compared with the value of projects under funding consideration.

According to the DWSRF program, states are required to prioritize the use of funds to projects that "address the most serious risk to human health; are necessary to ensure compliance with the requirements of the Safe Drinking Water Act; and assist utilities most in need, on a per household basis, according to state-determined affordability criteria." (USEPA 2014) Therefore, timing of project funding needs and the prioritization of the project based on the project's purpose are considerations that affect the suitability of SRF loans. In addition, the amortization period of government loans is usually a fixed term that is specified by the program and may be shorter than other forms of debt.

The CWSRF program is generally better funded than the DWSRF program, and "States have the flexibility to target resources to their particular environmental needs, including contaminated runoff from urban and agricultural areas, wetlands restoration, groundwater protection, brownfields remediation, estuary management, and wastewater treatment." (USEPA 2014)

The WIFIA program uses US treasury funds to fund larger (greater than $20 million) water, wastewater, and stormwater projects with loans at treasury rates, plus a small add-on for administrative costs. A five-year pilot WIFIA program was approved in 2014. The WIFIA program has several unique provisions. The WIFIA program:

- Is aimed at larger projects, in which eligible projects must cost at least $20 million ($5 million for communities serving no more than 25,000 people).
- Provides loan guarantees and direct loans at long-term Treasury rates. Projects must be deemed creditworthy, with loans repayable from a dedicated revenue source within 35 years of substantial project completion.
- Allows a private entity to secure a WIFIA loan with the sponsorship of the applicable public authority (see Public–Private Partnerships).
- Gives states the "right of first refusal" to fund any project with the SRF program.
- Limits WIFIA support of a project to 49 percent of the project's costs, with an overall limitation of 80 percent for all federal assistance in any project (with an exception for certain federally funded projects in Indian tribal communities), and provides that tax-exempt debt cannot be used to pay the nonfederal share of project costs. However, in any year the USEPA Administrator can use up to 25 percent of appropriated funds in projects exceeding the 49 percent limitation, up to the 80 percent federal maximum.
- Reserves 15 percent of appropriated funds each year for projects in communities with a population of no greater than 25,000. By June 1 of each year, any such reserved funds that have not been obligated shall be available for projects in communities of any size.
- Authorizes $20 million in the first year, which should support at least $200 million in loan guarantees or low-interest loans. The authorization level rises to $50 million in year five, which should support up to $1.65 billion in assistance, according to the US Office of Management and Budget. Appropriated funds achieve this significant leverage because they only have to cover the risk of WIFIA project defaults, and the history of default in water projects is only 0.04 percent, according to AWWA research (AWWA 2014).

Leases

Leases often provide additional flexibility to a utility and can be offered at taxable or tax-exempt rates. A lease at a taxable rate may be more expensive to a government-owned utility than using funds from other sources. However, if the utility has only a short-term need for specialized equipment, or if equipment is expected to become obsolete before it wears out, the utility may find it advantageous to lease the equipment, even at the higher rate, especially if the equipment can be returned or upgraded before reaching the end of its useful life.

The most obvious benefit of a lease is the low cost. Lessors can pass on a portion of their tax savings to the lessees, resulting in a lower cost for the lessee. A disadvantage of a lease is the complexity and initial cost in arranging such financing.

A lease in the form of a *certificate of participation* (COP) has many of the same advantages and disadvantages as a revenue bond. The cost to the utility to obtain funds through the use of a COP structure depends both on the credit rating of the utility and on how essential the leased facility is to the utility. A COP secured by essential facilities is more attractive to investors because it is generally secured only by rental payments on the leased facility.

Contributions

Contributed funds, from developers and/or new customers of the utility system, enable a utility to require that new users share in the full cost of developing utility service capability. These contributions can be made through the construction and dedication of assets or through payments known as *system development charges* or *capacity fees*. Contributions may be used to expand service into new areas or pay a proportional share of costs associated with upgrading facilities that will accommodate new users. In either case, contributions allow the utility to maintain lower costs than might otherwise be needed. Contributions may help postpone or reduce the amount of future rate increases. They may also pay for capacity reserved for growth, the cost of which existing customers may otherwise have paid for over a number of years. Through a utility's collection of contributed funds, the need is minimized for current customers to subsidize future expansion or rehabilitation of the water system.

The disadvantages of contributed funds include the inconsistent levels and timing of the funds receipt. A utility must have extra capacity in its water facilities prior to physically being able to serve new customers. Typically, capital investment precedes the receipt of contributed funds as capital facilities are constructed in anticipation of increased demands. Receipt of contributions usually depends on growth cycles and the demand for utility services, which can be inconsistent over time and difficult to precisely predict. Additional administration is often required as contributed funds are restricted and thus must be accounted for separately.

There are cases where utilities are allowed to use contributions (development fees) as revenue toward debt-service coverage. However with recent changes in growth patterns and accompanying changes in developer fee contributions, this revenue stream has, in some cases, proven to be unpredictable and unreliable, requiring utilities to raise rates and/or deplete reserves as this revenue stream has declined. Caution should be used when including contributed funds as revenue toward debt-service coverage.

Grants

Using grant funds, if available, generally helps utilities reduce the total costs of specific capital projects. Unlike other forms of funding options, grants are not required to be repaid by the utility. However, this form of funding is generally specific to certain types of projects, may require the utility to qualify for the grant, and is generally competitive. However, with partial grant funding, local governments may be inclined to oversize systems or facility capacity because the utility's money goes considerably farther. Because such oversized facilities may not work as efficiently, oversizing may result in higher operating expenditures and greater annual cash expenses. This could impact water rates until the excess capacity is absorbed.

Most grant funding requires increased reporting and strict compliance with grant provisions. This could be a disadvantage if the utility's needs change in the future. Change in use may require the utility to incur unforeseen expenditures at a future date or even require repayment of the grant itself. Few pure grant funding programs exist, and applications to programs that are currently available are highly competitive and need-based.

Tax Benefit Districts

Tax benefit districts provide funding through a special tax (generally ad valorem) in a specific defined area (the district) for those states that employ them, such as New York, Connecticut, Virginia, and Alaska. The tax benefit district aligns the cost of a project or projects with the customers receiving benefit from the projects. This differs from other forms of funding that are secured and repaid without considering whether the customers benefitted from the projects. In addition, customers in a tax benefit district can deduct the cost of this tax from their annual income taxes, unlike other forms of funding that are recovered through nondeductible rates and charges.

The disadvantage of using tax benefit districts pertains to the need to establish the district or to increase the tax limits for rehabilitation purposes by a majority vote of the customers and undeveloped properties that will benefit from the service. A considerable public relations effort by the utility may be necessary to convince property owners to establish a tax district. If needs are properly defined and corresponding benefits are carefully explained to property holders and responsible public officials, the initiative could be successful. If the vote fails, that could be perceived as a referendum against the utility for making improvements to its facilities in the area. However, a negative vote is generally a sign that the utility must employ some other means of financing the needed improvements.

For utilities that have historically financed improvements and rehabilitations to their systems through general schedules of rates and charges for service applicable to all customers, the use of tax-benefit districts will fundamentally change and potentially complicate financing. The use of multiple tax benefit districts may result in the total cost of utility service being different in each district. These differences may affect the pattern of customer growth in the service area. This in turn may affect the utilities' capital planning and ability to provide timely service to each district. Some districts may feel that their needs are ignored while other districts' needs are addressed.

Historically, utilities have elected to spread their costs to all customers to minimize observable differences in utilities' costs to serve the various parts of their service areas. This practice tends to minimize customers' publicly expressed dissatisfaction with utilities' rates and charges. Without the use of cost averaging during rate setting, some communities may become concerned that customers who can afford to pay will benefit by lower billings for utility service and those who cannot afford it will pay more than a perceived fair share.

Public–Private and Public–Public Partnerships

Public–private partnerships (PPPs) are a contractual arrangement whereby a private entity agrees to provide resources, shared or spread risks, and share rewards of investing in water or wastewater services. In such an arrangement, the partnership can provide a water or wastewater system greater efficiency, better access to capital, and improved compliance with a range of government regulations regarding the environment and workplace. The public's interests are assured through provisions in the contracts that provide for ongoing monitoring and oversight of the operation of a service or development of a facility. Public-private partnerships come in a variety of forms, and no two PPP projects are exactly alike. These partnerships can range from relatively short-term operation and maintenance of a facility or system to long-term design–build–finance–operate–maintain, in which essentially all of the services are transferred to the private partner.

Public–private partnerships in the United States can result in a higher cost of borrowing from the private entity when compared to tax-exempt debt that can be issued by a municipality. However, the higher cost of debt may be offset by the cost savings from a PPP arrangement.

The WIFIA program, as discussed in the Government Loans section, may result in greater opportunity for PPP arrangements. The WIFIA pilot program limits WIFIA funding to 49 percent of a project's cost, with tax exempt debt excluded from funding the remaining project balance. It is expected that WIFIA will encourage forms of private financing such as PPP arrangements.

Public–public partnerships are another innovative arrangement, in which two or more public authorities agree to cooperate to their mutual advantage. Such arrangements can range from the relatively simple, such as a concession agreement in which one entity assumes certain services, like billing, for the other entity; to more complicated, such as an agreement for the complete financing, operation, and maintenance of one system by another; to the outright merger of previously separate systems.

BOND STRUCTURE VARIATIONS

An issuer can increase the marketability of its bonds by using a variety of bond structures to address specific market conditions. Investment bankers continually try to satisfy both issuers and borrowers by developing new bond structures and instruments that offer

Table 3-1 Comparison of funding alternatives

Funding Alternative	Key Components	Advantages	Disadvantages	Conditions Favorable for this Alternative
GO bonds	Pledge of full faith and credit; tax-backed debt	Strongest security for investors; low debt interest and issue costs	Requires voter approval; legal debt capacity limitations	Healthy or growing tax base; no debt ceiling concerns; city healthier than enterprise
Revenue bonds	Pledge of enterprise revenues; provides a dedicated funding stream	Generally requires no voter approval; no draw on debt capacity or tax base; user fee schedules	Higher debt interest cost than GO bonds; more complex and costly sale than GO bonds	Enterprise healthier than city; city at or near debt ceiling for GO debt; relieves new tax pressures
Government loans	Loans available through federal or state programs such as WIFIA, SRF	Below-market interest rates; low issuance costs; few restrictive covenants	Competitive; selection may be based on project prioritization; inflexible debt amortization terms	Projects with long lead time before funds are needed, or that receive a high-priority ranking
Double-barrel or combination bonds	Pledge of revenues backed by pledge of full faith and credit	Additional security for enterprise shortcomings; revenue pledge may keep bonds out of debt limit	Requires voter approval; may have legal debt capacity limitations	Enterprise may not have a record of debt issuance or is in some distress; poor market interest in the debt
Tender-option or put bonds	Long-term bonds with tender option prior to maturity; issue has short-term debt characteristics	Takes advantage of lower short-term rates	Requires credit facility to ensure redemption funds; savings vary with market; more complex and costly	Best used during periods of steep interest yield curves
Floating rate or variable rate bonds	Rates linked to a market, treasury, or interest rates index; periodically revised; high–low limits set	Lower rates because issuer assumes full interest rate risk; popular in a volatile bond market	Costly to administer; uncertainty of future debt service payment levels; higher risk to issuer	Attractive investments when investors expect interest rates to rise

(Table continued on next page)

Table 3-1 Comparison of funding alternatives (continued)

Funding Alternative	Key Components	Advantages	Disadvantages	Conditions Favorable for this Alternative
Zero-coupon or capital appreciation bonds	Long-term bonds with deep discounts from par; no interest payment before maturity	Lower interest rates because issuer assumes full reinvestment risk; low administrative cost	Sinking fund must meet balloon payment at term; maximum debt approval may be difficult to obtain	Less prevalent since 1986 Federal Tax Act changes; check state restrictions on deep bond discounting
Original-issue discount bonds	Long-term bonds sold with low interest rates at discount below par	Lower interest costs to issuers; lower overall yield to maturity	Greater amount of bonds must be sold to realize same amount of proceeds	Attractive during periods when interest rates are thought to be declining
Bonds with municipal bond insurance	Purchase of bond insurance gives issuer a AAA bond rating	Low to medium credit ratings become AAA; lower interest expense	Insurance payment paid from bond proceeds; additional insurers' limits may be incurred	Investment-grade rating needed; insurance cost less than interest payment spread for credit ratings
Bonds with letter of credit (LOC)	Major banks to cover debt service payments in case of issuer default	Lower interest cost to issuer because of upgrade of rating	Expense to obtain an LOC; higher costs for administration	Need to raise ratings; banks' creditworthiness exceed that of issuer
Bond-, revenue-, tax-, and grant-anticipation notes	Short-term borrowing for less than one year; bank LOC is security	Lower interest costs to issuer	High issue/administrative costs; with interest rate rise, LOC may not renew at term	Note limited to 10%–15% of debt portfolio; used when interest rates are high and expected to drop
Commercial paper	Promissory note with maximum of 270-day term (average 30–45 days); rolled over continuously	Considerably lower interest costs to issuer; strong demand by funds and individual investors	Requires LOC to ensure liquidity; administration costly; LOC may not be renewed as rates rise	Note limited to 10%–15% of debt portfolio; used when interest rates are high and expected to drop

acceptable terms to both parties. Historically, such structuring included the development of variable-rate demand obligations (put bonds), floating-rate obligations, original-issue discount bonds, zero-coupon bonds, and capital appreciation bonds. The issuer should be aware that if it selects such variations to sell the bonds, it may be accepting more interest rate risk. The issuer should discuss alternative structures in depth with its financial advisor and ensure that they understand both the benefits and the risks that can be expected. Characteristics, advantages, and disadvantages of several of these structures and instruments are discussed in the following section and are summarized in Table 3-1.

Variable-Rate Demand Obligation

The most significant advantage of variable-rate demand obligations (tender-option or put bonds) is the interest savings, which can be substantial. Investors are willing to accept a lower rate on these instruments during a period of steep yield curves because the put option (ability to sell the obligation back to the trustee) provides a way to exit the bonds at par before maturity. A put option provides the bond holder with the right, but not an obligation, to sell the bonds within a certain time period for a specified price.

The amount of savings will, however, vary with market conditions, the credit rating of the issuer, and particular put-option features. In addition, the letter of credit (LOC) fee, any insurance premiums, and the cost of funds potentially borrowed against the LOC can offset potential savings.

Discount Bonds

Discount bonds have two primary advantages for the issuer: lower interest costs and a generally lower overall yield to maturity. There are a number of reasons investors are willing to accept a lower yield and lower interest income from discount bonds. In states where the face value of a discount bond can be used to meet pledging requirements, the same dollar amount can buy considerably more face value of discount bonds than conventional bonds. Similarly, individual investors can purchase a large par amount of discount bonds at a relatively low cost. Discount bonds are also advantageous during a period of active tax swapping, when they are in considerable demand.

Bonds with extreme discounts, such as zero-coupon or capital appreciation bonds (CABs) have additional advantages. First, with these instruments, the investor no longer has reinvestment risk, so these bonds are particularly attractive to investors during periods when interest rates are expected to decline and only lower-yielding reinvestment options are anticipated. Second, zero-coupon bonds appeal to a broad market because they are often purchased by individual investors, for whom a $5,000 bond might be prohibitively expensive, to provide money for future educational or retirement needs. Because these bonds are attractive to investors, many investors accept a lower yield to maturity than on a conventional bond. Another advantage to the issuer is that a zero-coupon bond incurs fewer administrative expenses, because there are no semi-annual interest payments.

CABs are distinct from traditional zero-coupon bonds because the investment return is considered to be in the form of compounded interest rather than the accreted original issue discount. For this reason, only the initial principal amount of a CAB would be counted against a municipal issuer's statutory debt limit, rather than the total par value, as in the case of a traditional zero-coupon bond.

A disadvantage of discount bonds for an issuer is that the issuer must sell a greater par amount of bonds to realize the same amount of bond proceeds. This could present problems if the issuer's legal debt authorization is limited. Another consideration for an issuer is that redeeming discount bonds early is expensive. Exercising call provisions at premium or even at par can raise the yield on the bonds considerably.

Bonds with extreme discounts also have disadvantages for issuers. The issuer must carefully structure appropriate sinking-fund provisions to ensure that requisite funds will be available at maturity to meet balloon payments due at that time. Also, some states and provinces may prohibit excessive discounting of bonds. In other instances, state/provincial or local debt limitations or maximum debt authorizations may inhibit the sale of capital appreciation or zero-coupon bonds. Further, it may be difficult to convince the public that issuing bonds with a par value many times greater than bond proceeds is appropriate.

Bonds with Municipal Bond Insurance

The municipal bond insurance market can be volatile, and at times this insurance may become difficult to obtain. Due to this volatility, the cost of municipal bond insurance can fluctuate and at times may be very expensive, making this enhancement cost prohibitive for many utilities. At a time when municipal bond insurance may not be available or be cost prohibitive, the utility may need to increase reserve funds. Nevertheless, there are advantages of using bond insurance or other credit enhancements. These enhancements help the utility achieve the highest possible credit rating, thus enabling a low-rated issuer to pay a lower interest rate than its own credit rating would normally allow. The use of insurance may also broaden the market for the bonds because some investors demand bond insurance or a minimum rating.

The disadvantages and risks associated with the use of municipal bond insurance include the front-end deduction of the entire insurance payment from bond proceeds, thereby reducing the amount of funds available for construction projects. In addition, bond insurance may not be available to all issuers. To spread their risk, insurers may have limited underwriting capacity in certain geographic areas or with regard to particular types of bond issues. Insurers also frequently impose restrictions on bond issues they underwrite, including requiring reserve funds, high coverage requirements, or enforceable rate covenants.

Bonds with Letter of Credit

The advantages of using an LOC to enhance the rating on a bond are the same as those previously discussed for bond insurance. LOC support can produce considerable interest savings for an issuer. Also, LOCs may be a requisite component of such financing techniques as variable-rate demand bonds and commercial paper (CP).

A disadvantage of using LOCs is that few banks provide commitments beyond 5 to 10 years, so most issues supported by LOC have relatively short maturities or are structured to provide for retirement of the debt before expiration of the LOC. Bonds issued for shorter periods tend to have higher annual debt service payments than might otherwise be the case, resulting in a greater financial effect on utility financing. Also, the cost of securing an LOC can add measurably to the ongoing expenses of a bond issue. The fee charged is generally payable on an annual basis and represents a fixed percentage of the amount of bond funds secured by the LOC. In addition, utility administrative costs may increase because of requirements to file regular financial reports with the bank issuing the LOC.

Furthermore, utilities should consider rollover risk when evaluating bonds with letters of credit. A utility may be unable to find a bank that will provide an LOC when the current one expires. If not, the utility may be forced to refinance debt at higher interest rates upon expiration of the LOC, and/or a bank may require termination of the LOC if the utility cannot meet financial covenants or is downgraded, forcing the utility into the market when its credit condition has weakened.

Bond-, Tax-, Grant-, and Revenue-Anticipation Notes

The major advantage of issuing bond-anticipation notes (BANs), tax-anticipation notes (TANs), grant-anticipation notes (GANs), or revenue-anticipation notes (RANs) instead of long-term bonds is the lower interest rate. At risk, however, is the issuer's ability to enter the long-term bond market at rates less than current interest rate levels when the notes mature. If rates do not decline measurably, future financing costs may increase. Also, if legislation permits, the temptation to roll over the outstanding notes with another short-term issue may overburden the municipality with short-term debt.

Commercial Paper (and Tax-Exempt CP)

The primary advantage of commercial paper for an issuer is the extremely low interest rate. Because of its short maturities, superior credit quality, and ready liquidity, CP is a prime investment instrument for money market funds. Because of the strong demand for short-term securities by funds and individual investors, the interest rates on CP are the lowest of all instruments.

There are, however, a number of disadvantages to CP financing. The initial start-up and operating costs of such a program may be greater than those of alternative short-term borrowing methods. They include the cost of the backup credit facility, bond counsel fees, and transaction costs associated with administering the program. Consequently, $50 million is the generally accepted minimum issue size. Once CP has been issued, the recurring refunding and sale process that gives CP the characteristics of a continuous securities offering requires daily staff commitment. Issuers also assume all the risks of short-term borrowing when they issue CP. Depending on market conditions, the maturing issues may not be refinanced with new commercial paper. Another consideration for tax-exempt issuers is that if the backup line is drawn on, the new loan will probably carry a taxable interest rate.

SUMMARY

When evaluating alternative methods of financing capital expenditures, a utility will need to consider the legal and financial environments in which it operates, as well as the advantages and disadvantages of the various forms of financing that are available. In making a selection, the utility will want to consider not only its financial capacity but also the public acceptability of the various alternatives. How the various alternatives will affect the utility's future financial flexibility and how they fit within its long-term financial plan are also important considerations.

REFERENCE

US Environmental Protection Agency (USEPA). 2014. How the CWSRF Program Works. water.epa.gov/grants_funding/cwsrf/basics.cfm. Accessed May 8, 2014.

This page intentionally blank.

Chapter 4

Preparing the Utility for Financing

Before attempting to secure financing from outside sources, water utility managers can proactively take multiple steps to help the utility look its best to third parties such as credit rating agencies and lending institutions. A utility's credit rating is an assessment of financial risk and a reflection of the actions of utility management. Management quality is a key aspect for credit assessment and is reflected in the utility's financial operations and budgeting. For example, management quality may be reflected in the willingness and ability to raise rates; the effectiveness of staffing practices; capital improvement planning; construction management; and the organization's ability to meet the demands of changing regulatory, political, and economic environments (Moody's 1999). Management's decisions and initiatives, subject to the oversight and strategic direction of the governing body, will determine the utility's long-term stability and financial viability.

Effective management practices can help mitigate the operational and financial risks associated with running the utility business. A lower risk profile can translate to the utility receiving higher credit ratings and lower interest rates on utility debt financing. Lower interest rates can help the utility maintain affordable user rates for the long-term benefit of its customers.

This chapter discusses certain utility best management practices and the characteristics of highly-rated or financially strong utilities, especially the factors over which utility managers have control. While the chapter focuses on publicly owned utilities, some of the practices discussed are also applicable to private utilities. By following these guidelines, utility management can position the utility to receive more favorable credit ratings. Many of these practices require a significant amount of time to implement and, as such, utility managers must plan ahead and adopt a long-term perspective when making operational and financial decisions for the utility.

HOW THIRD PARTIES EVALUATE THE UTILITY BUSINESS

A water utility is a capital-intensive business, meaning that a large amount of capital investment is required in order to provide service. For most local governments operating a utility, the utility's operations are the largest government business and are accounted for in an enterprise fund. Under generally accepted accounting principles, enterprise funds are used to report operations that are financed and operated like private businesses, where the intent is to recover expenditures completely or partially from income collected through user rates.

Credit rating agencies and other organizations such as AWWA provide guidance for effectively managing the utility business or enterprise. One reference published by AWWA is *Financial Management for Water Utilities: Principles of Finance, Accounting and Management Controls.*

The three major credit rating agencies are Moody's Investors Service, Standard & Poor's Ratings Services (S&P), and Fitch Ratings. These companies publish utility industry outlooks, financial medians, and criteria for rating utilities, either in print or online. It is helpful for utility managers to be on the distribution list for such publications or to have access to them. These references are good information sources for industry trends, performance measures, and the issues affecting utility ratings. The three rating agencies have also published their general rating considerations for water and sewer revenue bonds. Table 4-1 lists the website, publications, and rating criteria for these agencies. Appendix A discusses the factors used by rating agencies to assess the risk of owning a utility's securities.

Utility managers can control or influence many of the rating factors. The next section discusses the management practices of highly rated or financially strong utilities. If these practices are followed, a utility should be looked upon favorably not only by rating agencies and potential investors, but also by lending institutions and federal/state loan and grant programs.

MANAGEMENT PRACTICES OF HIGHLY RATED UTILITIES

The remainder of this chapter is dedicated to discussing the management practices of highly rated or financially strong utilities, which include the following:

1. Maximize stability through long-term operational and financial planning.
   a. Ensure that the utility has an adequate water supply.
   b. Establish business principles.
   c. Adopt a long-term business plan.
   d. Adopt financial policies and performance measures.
   e. Adopt a formal debt-management policy.
2. Meet financial targets recognized as prudent by the utility industry and rating agencies.
3. Maintain healthy cash balances.
4. Ensure that the financial plan enables the utility to maintain affordable user rates.
   a. Develop a realistic capital plan.
   b. Evaluate operations and operating expenses to achieve efficiency and effectiveness.

Table 4-1 Water and wastewater revenue bond rating agency detail

Rating Agency	Publications	Key Bond Rating Factors
Fitch Ratings www.fitchratings.com	*Water and Sewer Revenue Bond Rating Guidelines* *Rating Criteria for Infrastructure and Project Finance* *Median Ratios for Water and Sewer Revenue Bonds* (published annually)	Fitch 10 Cs of Water / Sewer Revenue Bond Analysis* Crew (management) Community characteristics Customer growth and concentration Capacity Compliance with environmental laws and regulations Capital demands and debt policies Covenants Charges and rate affordability Coverage and financial performance Cash and balance sheet considerations (Fitch 2007)
Moody's Investors Service www.moodys.com	*Analytical Framework for Water and Sewer System Ratings* *Enterprise Performance Medians for Water and Sewer* (published annually) *Moody's on Revenue Bonds: The Fundamentals of Revenue Bond Credit Analysis*	Governance and Management Quality Regulatory Risk Construction Risk System Size and Assessment Base Local Economy and Customer Base Strategic Focus System Demand and Capacity Maintenance of Assets Regulatory Compliance Rates, Rate Structures, and Ratemaking Flexibility Liquidity (Moody's 1999)
Standard & Poor's Ratings Services www.standardandpoors.com	*Public Finance Criteria Book* *U.S. Public Finance: Key Water and Sewer Utility Credit Ratio Ranges*	Management Economic Considerations Financial Data/Capital Improvement Plan Rate Criteria Operational Characteristics Legal Provisions (Standard & Poor's 2009)

* Fitch 10 Cs are a subset of Fitch's four general areas of revenue-supported rating criteria: governance and management, financial profile, debt profile, and operating profile.

   c. Adopt an automatic rate-adjustment clause.
   d. Consider refinancing existing debt.

5. Adopt a rate structure with significant cost recovery through base charges to promote revenue stability.
6. Minimize dependence on fees charged to new growth to pay debt.
7. Demonstrate support from regulators (i.e., governing bodies).
   a. Adopt rate increases needed during the forecast period.
   b. Develop a program for public outreach and stakeholder involvement.
8. Remain in compliance with applicable federal, state, and local regulations.
   a. Ensure that the utility's nonrevenue water percentage is within industry norms.
9. Limit nonutility-related transfers to the general fund.
10. Implement regular reporting and monitoring systems.
11. Have collection policies and practices that regularly track receipts and enforce penalties against late payers.
12. Incorporate best financial management practices into the bond resolution, ordinance, or indenture authorizing the issuance of bonds.
    a. Require renewal and replacement funding.
    b. Have a closed-loop system.
13. Plan for contingencies.
14. Recognize additional bonds test when developing the financial plan.
15. Be prepared to provide tours of utility facilities.

Each of these items will now be discussed in detail.

Maximize Stability Through Long-Term Planning

Utility management should establish and have the utility regulators (e.g., governing bodies) officially adopt operational and financial policies as well as a long-term business plan (such as a master plan); comprehensive development/growth management plan; and debt-management plan. Long-term planning is essential for establishing and maintaining a strong utility business foundation. Utilities should also maintain written documentation on such policies and plans, as rating agencies often request copies of such documentation. Ideally, utility managers should be able to show third-party evaluators that the utility has established the policies and plans and has a consistent track record (history) of following them.

Ensure an adequate water supply. Water is becoming an increasingly scarce resource in many areas across North America, and a utility's water supply planning must ensure that the utility's water demands can be met for the foreseeable future. Water supply planning by the utility management can include:

- Securing the necessary water withdrawal permits from water supply regulatory agencies.

- Negotiating long-term contracts or inter-local agreements to purchase water from other utilities or entities, or to jointly share in regional facilities with other utilities or entities.
- Expanding the reclaimed water system to provide for the irrigation needs of system customers and reduce potable water demands.
- Designing the water rate structure to encourage water conservation, thereby reducing water demand. The utility can also implement a public education program to promote water conservation efforts.

Establish business principles. Business principles provide the basis for long-term decision making for the utility and should be aligned with stakeholder expectations. Examples of business principles are:

- The utility is customer-focused, and customer satisfaction is the highest priority.
- The utility will maintain financial creditworthiness and meet financial benchmarks recognized as prudent by credit rating agencies and the utility industry.
- The utility will establish affordable rates to recover the full cost of providing service.
- The utility, to the extent practical, will establish cost-recovery practices and rates such that growth will pay for growth; existing customers will not subsidize growth of the utility.
- The utility will comply with all requirements of federal, state, and local regulatory agencies (e.g., departments of environmental protection and health, water management districts) in order to meet public health and safety requirements.
- The utility will take the necessary steps to ensure that service is uninterrupted, including the (1) implementation of timely renewals and replacements to reduce the risk of system failures, and (2) preparation for emergencies (e.g., adverse weather conditions such as hurricanes).

It is appropriate for the utility to have a mission, core values, vision, goals and objectives, as discussed in chapter 1, and to specify levels of service to its customers.

Adopt a long-term business plan. The need for a utility business plan is driven by the fact that a utility

- provides a necessary service,
- has major expenditure needs that must be funded by the community,
- has long-lived assets, some of which may last 50 to 100 years,
- typically funds a large amount of its capital needs through long-term financing,
- needs to avoid financial surprises, and
- should position itself to lower borrowing costs through long-term planning.

The utility business plan should focus on both the operational and financial aspects of running the utility business. Some utilities have developed comprehensive business plans similar to what might be prepared for a private company, while other utilities have multiple long-term planning documents that function as components of the overall business plan, such as a master plan.

Many local governments require their utilities to have current master plans, and laws often specify how often such plans must be updated. It is important to modify expansion and sustainability plans based on the latest service-area population projections and other economic, environmental, and regulatory factors affecting the utility service area, as well as infrastructure needs. The master plans should be developed by professional engineers and updated periodically (e.g., every two to three years) to reflect current costs and service area needs. A utility's development/growth management plan should enhance diversification and focus on generating solid customer and income growth.

Adopt financial policies and performance measures. Many factors can cause financial volatility, such as water supply variations, water demand due to weather, and economic cycles. The highest-rated utilities establish financial goals or targets and consistently develop rates and budgets that comply with such targets.

At a minimum, financial policies should contain targets or requirements for debt-service coverage and cash reserves to ensure appropriate financial margins and to be prepared for the cyclicality of cash flows. According to the *Water and Sewer Revenue Bond Rating Guidelines* published by Fitch Ratings, "For utilities in the most stable operating environments with a suitably diverse and healthy service area economy, 1.5× annual coverage, with consistently maintained unrestricted financial liquidity of at least 90 days of operating revenues, could be sufficient for 'AA-' or higher ratings. For utilities with substantial growth, compliance demands, or significant annual volatility in revenues or expenditures, greater financial flexibility may be necessary" (Fitch 2007).

Utilities should strive to exceed the minimum debt-service coverage required in their bond or loan documents, as adequate excess revenues can allow the system to cash-fund a greater proportion of capital needs. The *all-in debt-service coverage* is a key ratio used by credit-rating agencies and is defined as "annual revenues available for debt service divided by the total annual debt service." *Total annual debt service* refers to senior and subordinate lien debt service combined. *Annual revenues available for debt service,* which are often the utility's net revenues, are typically derived from the definitions of revenues and operating expenses in the utility's authorizing bond ordinance, resolution, or loan agreement. The most common formula to calculate such revenues is gross revenues less operating expenses, which is usually derived as follows:

User Rate Revenues + Other Operating Revenues + Unrestricted Interest Income[1] = Gross Revenues

Gross Revenues – Operating Expenses[2] = Net Revenues or Annual Revenues Available for Debt Service

The target debt service may depend upon the requirements of the underwriter, the rating agencies, and the investors. A target all-in debt-service coverage of 150 percent is usually reasonable, as is a working capital target of 90 days of rate revenue. However, lending institutions and the financial market may demand greater coverage and/or liquidity depending on a utility's overall capital needs, service area size and requirements, potential impact of new regulations, or susceptibility to catastrophic events. In 2012, the median all-in annual debt-service coverage excluding connection fees for utilities rated "AAA"

[1] Interest income from accounts restricted by law or other financial covenants, such as interest earnings from system development charge funds and construction funds established for bond proceeds, is typically not included in the coverage calculation.

[2] The definition of *operating expenses* or the *cost of operation and maintenance* (O&M) in most bond ordinances and resolutions excludes noncash expenditures such as depreciation and amortization.

by Fitch Ratings was 220 percent, while the median for AA-rated and A-rated utilities was 180 percent and 140 percent, respectively (Fitch 2012). For liquidity, Fitch publishes medians for a key ratio called *days cash on hand*—see the sections, Meet Prudent Financial Targets and Maintain Healthy Cash Balances, for more information about this concept. It is important to recognize the definition of *median*; half of the utilities with a given rating had a debt-service coverage above the median, while the coverage for the other half was below the median.

It is also appropriate to establish as a financial policy a methodology for capital reinvestment for ongoing renewals, replacements, upgrades, and betterments. Regardless of the expenditure requirements, it is appropriate to make annual deposits or expenditures for capital reinvestment. Ideally, such deposits should be based on a comprehensive asset management plan. However, many utilities base the deposits on a percentage of revenues. Capital reinvestment within the utility industry has historically been very low. In its *2013 Report Card for America's Infrastructure*, the American Society of Civil Engineers (ASCE) assigned a grade of "D" to America's drinking water systems, citing billions of dollars of annual funding shortfalls to replace aging facilities near the end of their useful lives and to comply with existing future federal water regulations (ASCE 2013).

In 2012, AWWA published its report entitled *Buried No Longer: Confronting America's Water Infrastructure Challenge*, which noted, "Much of our drinking water infrastructure, the more than one million miles of pipes beneath our streets, is nearing the end of its useful life and approaching the age at which it needs to be replaced. Moreover, our shifting population brings significant growth to some areas of the country, requiring larger pipe networks to provide water service. … restoring existing water systems as they reach the end of their useful lives and expanding them to serve a growing population will cost at least $1 trillion over the next 25 years, if we are to maintain current levels of water service" (AWWA 2012).

By annually funding a realistic allowance for ongoing capital replacement, utility rates over the long term are generally lower and do not fluctuate wildly with capital funding requirements (which often have a "lumpy" expenditure curve).

Some utilities have financial policies that require transfers for pay-as-you-go capital funding be at least as much as the annual depreciation expense. It is important to note that, in establishing rates to generate sufficient revenue to fund such transfers (as well as to fund operating expenses and debt service), a utility can more easily meet its debt-service coverage requirements since the transfers are typically "below-the-line" expenditure requirements.

Adopt a formal debt-management policy. Utilities should develop debt-affordability guidelines or models. The affordability analysis involves a complete review of existing and proposed debt and how the future financial profile will be affected; debt per customer and per capita should be considered. Debt management policies often address:

- What types of projects can be funded through debt issuance.
- Amortization periods/debt terms. The term of the debt should not exceed the service lives of the combined assets being financed.
- Criteria for when refunding of bonds is allowed.
- Exposure to variable-rate debt, swaps, and other contingent liabilities.

Debt management also involves exploring opportunities to secure grants and low-cost financing alternatives.

Table 4-2 Some key financial ratios used by credit rating agencies

Financial Ratio	Definition
All-in debt-service coverage	Revenues available for debt service divided by total debt service (senior plus subordinate lien)
Senior lien debt-service coverage	Revenues available for debt service divided by senior lien debt service
Days cash on hand	Unrestricted cash and investments divided by operating expenditures minus depreciation, divided by 365
Working capital days of rate revenue	Working capital cash balance divided by operating revenue, divided by 365
Operating margin	Operating revenues minus operating expenditures, divided by operating revenues
Outstanding long-term debt per customer	Principal amount of debt outstanding divided by the number of customers
Outstanding long-term debt per capita	Principal amount of debt outstanding divided by the number of persons in the utility service area.
	Both current and projected debt per customer and per capita are considered.
Debt ratio	Net funded debt divided by the sum of net fixed assets plus net working capital
Free cash as percent of depreciation	Current surplus revenues after payment of operating expenses, debt service, and operating transfers out divided by current year depreciation

Meet prudent financial targets. When evaluating a utility, credit rating agencies look at a number of financial ratios that demonstrate the financial strength of the utility (Table 4-2). A utility manager's objective should be market access at reasonable cost, not necessarily maintenance of a particular rating. Note that key financial ratios and rating criteria are subject to change as the utility industry continues to evolve and rating agencies adjust their analytical framework accordingly. These ratios signify the way in which a utility handles its financial affairs in a number of ways.

All-in debt-service coverage. This ratio measures the financial margin to meet current total annual debt service with current revenues available for debt service. Since the ratio considers both senior lien and subordinate lien debt obligations, it is a better indicator of the utility's overall ability to repay its debt than the senior lien debt-service coverage alone. System development charges (sometimes called impact fees, capacity charges, connection fees, and other names) are often not included as revenues in the calculation of the all-in debt-service coverage, even if such charges are pledged toward the repayment of the debt. Collection of these growth-related fees depends on the utility's development activity, and hence is not as stable an income stream as user rates. As previously mentioned in this chapter, a target all-in debt-service coverage of 150 percent is usually considered reasonable by rating agencies.

Senior lien debt-service coverage. This ratio measures the financial margin to meet current annual senior lien debt service with current revenues available for debt service. A ratio of less than the legal debt service requirement as specified by the utility's rate covenant can indicate a technical default by the utility that may mandate further action, such as a review of the appropriateness of the current rates.

Days cash on hand. This ratio measures the financial flexibility to pay near-term financial obligations. The utility's available cash is a very important rating consideration, as is discussed in the section, Maintain Healthy Cash Balances. According to Fitch Ratings, utilities with stronger financial profiles might have days cash on hand equal to one year or more (Fitch 2012). However, ultimately, a utility's cash balance levels and targets may also be influenced by the utility's long-term capital investment program, and will fluctuate based on the mix and timing of its debt issues.

Working capital days of rate revenue. This ratio also measures the financial flexibility to pay near-term financial obligations. As previously mentioned in this chapter, a working capital target of 90 days of rate revenue is usually reasonable.

Operating margin. This ratio measures the financial margin to pay operating expenses. It is the inverse of the operating ratio. A high operating margin can indicate that the utility is funding a significant amount of expenditures (e.g., debt service, pay-as-you-go capital, transfers to the general fund) other than operating expenses. However, a low operating margin may signify that the utility has little debt, may not be investing in capital assets, or may not be performing necessary renewals and replacements.

Outstanding long-term debt per customer. These ratios measure the debt burden for ratepayers. Utilities with stronger financial profiles might have an existing and five-year projected debt per customer of $1,500 or less, and/or an existing and five-year projected debt per capita of $500 or less (Fitch 2012). (These amounts are in 2013 dollars and should be adjusted for inflation each year.)

Debt ratio. This ratio measures the debt leverage of assets and can indicate the system's capacity for additional debt. A ratio of 0 percent means that the system has no debt outstanding, while a 100 percent ratio means that the assets are 100 percent debt-financed. Utilities with stronger financial profiles might have a debt ratio less than 50 percent. However, the utility life cycle significantly affects the need for debt financing. Growth-oriented utilities that must build infrastructure in advance of new growth may be more reliant on debt financing than mature utilities with high capacity utilization (e.g., in built-out communities).

Free cash as percent of depreciation. This ratio measures the annual financial capacity to maintain existing infrastructure at current level of service from existing cash flows. Utilities with stronger financial profiles might have a free cash as percent of depreciation ratio equal to 100 percent or greater. In determining the "free cash," rating agencies may subtract from surplus revenues any transfers out of the utility system, such as transfers to the general fund.

Maintain Healthy Cash Balances

Maintaining healthy cash reserves is possibly the most effective practice that a utility can use to enhance its credit rating (Larkin 2005). For a water utility, there can be variances in cash flow due to seasonal demands, the amount of precipitation, or economic and customer trends. As such, the utility should have a reasonable amount of cash and cash equivalents for working capital.

The utility's cash balance policy should include working capital/operating reserves (equal to at least 90 days of rate revenues or higher) and a capital replacement fund. Other cash reserves should be established based on the specifics of the utility and may include allowances for rate stabilization, anticipated changes in regulations, alternative water resource supply funding, and emergencies/catastrophic events. In 2012, the median days cash on hand for utilities rated "AAA" by Fitch Ratings was 427, while the median for AA-rated and A-rated utilities was 418 and 285, respectively (Fitch 2012). Note, however, that rating agency cash-balance suggestions can vary. For example, Standard & Poor's published cash ratio targets are lower than Fitch (e.g., S&P suggests that 30 to 60 days of cash is "adequate," 60 to 120 days is "good," and greater than 120 days is "strong").

A rate stabilization fund is a debt-service coverage management tool created by many authorizing debt documents. During fiscal years in which there is strong coverage that exceeds the coverage target, utilities can make deposits to the rate stabilization fund, thereby lowering the revenues recognized for the year and decreasing the coverage. Any deposits to the fund represent deferred revenues, while any withdrawals from the fund are recognized as revenues. During a year when there may be difficulties in meeting the coverage requirement, a utility can make withdrawals from the fund to increase revenues and improve coverage. Using a rate stabilization fund can provide the utility with additional time to phase in rate increases to build revenues to the level required to meet coverage targets on a sustainable basis.

Plan Financially for Affordable User Rates

Local economic and demographic factors affect a community's financial capacity to support and pay for drinking water service. Fitch Ratings has indicated that it considers rates for combined water and wastewater service that are higher than 2 percent of the median household income (MHI)—or 1 percent higher than MHI for an individual water or wastewater utility—to be financially burdensome (Fitch 2012). The US Environmental Protection Agency (USEPA) considers rates that are greater than 2 percent of MHI for an individual water or wastewater utility may have a high financial impact on customers (USEPA 1997).

Utilities with a stronger financial profile might have residential charges for combined water and wastewater service that are less than or equal to 1.2 percent of MHI, or less than or equal to 0.6 percent for an individual water or wastewater utility (Fitch 2012). The USEPA considers rates for an individual utility that are lower than 1 percent of MHI are likely to have a low financial impact on customers (USEPA 1997). Utility managers can help to keep rates affordable by considering the following guidelines when developing the utility's financial plan.

Develop a realistic capital plan. The need to fund the utility's capital program is often the most significant driver for the need for rate adjustments. Utility managers must be able to explain the purpose of the capital program and have good total project cost estimates and capital spending (drawdown) projections. Because utilities must have the funds available prior to the appropriation and encumbrance of funds, understanding the total amount and the timing of funds needed for the plan is extremely important.

Developing a capital project prioritization ranking system to help isolate critical projects from those that could be deferred can help determine the allocation of limited funds. This also helps the utility manager to explain the project need to financial managers responsible for funding, the elected officials responsible for approval, and the stakeholders who pay the ongoing rates for utility service. Criteria to evaluate the projects could include—in order of priority—public health and safety (the project addresses working and public health/safety issues); regulatory compliance (the project promotes regulatory and contractual compliance); system reliability (the project reduces the risk and consequences of asset failure and addresses other attributes of asset management [redundancy]); community/customer benefit (the project addresses service issues such as pressure, taste and odor, and customer satisfaction); and sustainability (the project will result in long-term cost efficiencies and has environmental benefits).

Because of regulatory drivers, some utilities may not be able to develop realistic capital plans, especially if regular reinvestment in the system has not occurred in the past. In this case, the utilities should strive to adopt as many of the other best management practices as possible to prepare the utility for financing while gradually addressing the reinvestment needs of the system.

Rating agencies view favorably utilities that regularly consult with community development officials, regional and local planners, and demographers to predict and potentially limit infrastructure needs related to population and business growth. When evaluating capacity expansions, historical and projected capacity utilization percentages should be considered. This analysis should really focus on the number of equivalent residential connections (ERCs) served and reserved by development (identified separately) and the level of service (gallons per day of capacity) assigned to each ERC. An ERC, sometimes referred to as an *equivalent residential unit* and other terms, is a means of expressing different levels of customer demand for water in terms of an equivalent number of residential demand accounts.

The capacity planning process represents a balancing of risks. The risk from an engineering standpoint is that the enterprise would not have sufficient capacity to meet the service area needs. As such, in the planning of future capacity, the most conservative approach is to use a higher population or level of service estimate so that sufficient capacity would exist if such growth trend or possible change in use due an extraordinary event (e.g., weather) occurred. From a financial perspective, the growth expectation should be lower so that the enterprise can have reasonable assurance that revenues will be sufficient to meet expenditure needs. The key risk from a financial standpoint is that if the enterprise were to build too much capacity too far in advance of new growth, user rates would need to be increased to pay for the underutilized or unused capacity (e.g., debt service on the capital costs of the capacity and fixed operating expenses to maintain such capacity). Unused capacity is the most expensive capacity of a utility. Capacity utilization is considered by rating agencies when evaluating a utility's creditworthiness.

If a public utility is considering bond financing for a portion of its capital program, Internal Revenue Service (IRS) arbitrage rules (Internal Revenue Code 148) generally require that all bond proceeds must be spent within three years. The arbitrage rules were established so that local governments could not issue tax-exempt bonds, invest the proceeds in taxable securities, and make a profit. When sizing a debt issue, it is very important to carefully assess how much of the capital program could physically be executed during the three-year timeframe. Issuing too much debt can put pressure on the utility to raise user rates to repay the debt, and hence create an unnecessary financial burden on existing ratepayers.

When considering funding options for the capital program, the utility should first consider whether any capital projects would be eligible to be funded through grants or any other cost-free options. For all projects that must be funded through utility revenues, there should be a balance between debt financing and pay-as-you-go capital funding. Debt financing is most appropriate for assets with longer service lives (e.g., 15 years or more), asset replacements that are nonrecurring, or for major units of property (generally higher cost asset). Pay-as-you-go capital funding is appropriate for routine capital expenditures, annual renewals/replacements, and assets with shorter service lives or costs that are not excessive (minor units of property). A balanced debt-financed versus pay-as-you-go funding program can help keep user rates affordable.

It is appropriate to compare the annual transfers for pay-as-you-go capital funding with the utility's annual depreciation expense, as rating agencies will use this relationship as a measurement of the amount of ongoing system maintenance and replacements being performed. For aging utilities, annual capital spending that is below the amount of annual depreciated assets on a prolonged basis could indicate that substantial upgrades may be required in the near term to maintain system condition and regulatory compliance, and pressure to raise rates could increase.

Chapter 1 discusses the capital planning process, and chapter 2 discusses capital financing alternatives in further detail.

Evaluate expenses to achieve efficiency and effectiveness. Efficiency and effectiveness are two common objectives in utility management. Chapter 2 discusses actions that utilities can take to avoid and/or reduce costs. Many factors over which the utility has no control affect the costs of operations. Factors such as rate increases from electric power providers, chemical costs from suppliers, changes in fuel and oil prices resulting from global market changes, competition for scarce resources such as concrete and building materials, and union contracts that stipulate or guarantee a certain percentage salary increase each year may result in cost increases that are greater than changes in the consumer price index (CPI). Increased regulations and standards for water and wastewater treatment are other significant factors that could immediately affect the cost of operations. Although utilities attempt to manage the cost of operations, a significant amount of the expenses is uncontrollable and should be passed on to consumers. For the expenses that utility managers can control, it is important to identify any operating cost reductions that could be implemented without sacrificing the level of service approved by the governing bodies.

One important annual operating expense relates to the funding of pensions and other postemployment benefits (OPEB) obligations, which can represent substantial long-term risks to governments and investor-owned utilities, depending upon the employee benefits that are offered. In many cases, large gaps exist between the amount of assets available to fund the obligations and the actual obligations. Utility managers often face the choice of either making the full annual required contribution or risking a larger unfunded liability. Some utilities have tried to contain the growth of existing liabilities by reducing benefits for new hires. The utility should be prepared to discuss these issues with rating agencies and other third-party evaluators. Credit rating agencies have indicated that utilities with substantial unfunded pension and OPEB liabilities may be subject to credit downgrades, given the financial risk associated with having to fund such obligations in the future.

Adopt an automatic rate adjustment clause. The utility should consider adopting a rate indexing clause to annually adjust rates automatically for inflation or to pass through increased expenses (e.g., purchased water and wastewater services, electricity) without a formal hearing. Such a clause can provide rating agencies more assurance that rates will produce operating margins to keep up with inflation (in the absence of a rate study recommendation). The national and regional CPI published by the US Bureau of Labor Statistics (www.bls.gov), including the Water and Sewerage Maintenance Index (a component of the national CPI), are commonly used for rate indexing. Another common index used in the utility industry for automatic rate adjustments is the Gross Domestic Product Implicit Price Deflator Index published by the US Bureau of Economic Analysis (www.bea.gov).

Consider refinancing existing debt. When developing a financial plan, the utility manager should always consider whether interest rates, current market conditions, and bond provisions might enable the utility to save money through reduced interest costs by refinancing a portion of its existing outstanding indebtedness. If the utility has retained a financial advisor, the financial advisor should be able to provide guidance on refinancing opportunities.

Achieve cost recovery through base charges. Water sales and the corresponding revenues attached to such sales are subject to volatility due to conservation, economic factors, and other issues, such as weather and season. While many objectives need to be considered when developing service rates, revenue stability is an objective valued by rating agencies because it is directly linked to financial risk. This financial risk occurs because a significant portion of a utility's costs are fixed but a large portion of the revenues is generated through volumetric rates based on water consumption, which can vary due to weather and economic conditions. Fitch Ratings views favorably utilities whose base or fixed charges generate revenues that represent more than 30 percent of the total revenue stream (Fitch 2012).

Minimize Dependence on Fees for New Growth

Sometimes utilities issue debt to fund growth-related projects for an enterprise and then expect fees charged to new growth, such as system development charges, to pay the annual debt service. While this strategy may result in lower user rates during periods of high growth, collections of fees from new growth are likely to diminish during economic downturns, requiring the debt service to be paid with user rate revenues. In this situation, significant additional user rate increases may be required to meet rate covenants. To minimize financial risk, the utility should minimize its dependence on fees charged to new growth to pay debt service. User rates should be established at levels to meet all debt coverage requirements and targets.

Utilities should calculate an expansion percentage for each debt issue, as well as a weighted expansion percentage for all debt issues combined. The expansion percentage can be derived by reviewing the capital projects financed (or anticipated to be financed) with the debt proceeds and determining which ones are allocable to growth.

Many utilities pledge system development charges toward the repayment of debt. However, since system development charges by law can normally only be used to pay for expansion-related debt (or expansion-related capital projects), the expansion-related portion of the debt service should be calculated so that the amount of system development charges legally available to pay the debt service can be determined. Many debt-service coverage tests only allow the legally available system development charges—the lesser of (1) the total system development charges collected during the year (including interest earned on such fees); and (2) the total expansion-related debt service—to be included in the coverage calculation. Some bond resolutions, ordinances, indentures, and agreements require a utility to document and/or certify the calculated expansion percentages as part of the bond closing process.

Statutory and case law mandates certain practices regarding the implementation and use of system development charges (e.g., collections and expenditures pertaining to each system development charge must be accounted for in a separate fund). Utility managers should ensure that the utility's financial practices fully comply with all legal requirements for these fees.

Demonstrate Support from Regulators

Adopt rate increases needed during the forecast period. Rate increases should be considered by regulators and the public to be an integral part of the utility's business plan centered on utility best management and financial practices. These practices are necessary to promote effective utility management as it relates to operational efficiency, financial creditworthiness, and compliance with performance standards recognized by the utility industry and credit rating agencies. Such practices also help to ensure long-term rate sustainability/affordability. Utility decision making on rates should consider the long-term interests of customers from both an operational and a financial perspective.

One element of financial risk is whether governing bodies adjust rates when necessary. If regulators adopt the multiyear rate implementation plan before the additional debt is incurred, some of the risk is mitigated.

Develop a program for public outreach and stakeholder involvement. With a proactive education and information effort, routine or ongoing briefings to regulators, and regular meetings with the community, the utility can build support from its stakeholders. If regulators are impressed, they may be more inclined to raise user rates, and if customers are impressed, they may have less opposition to rate adjustments.

Some states and local governments have enacted laws that mandate the involvement of stakeholders other than the governing body in the utility's operational and financial planning. For example, some laws (e.g., statutes of certain states) require an advisory committee consisting of members from the development/building community to be involved in the utility's growth planning and creation of the capital improvement program.

Remain in Compliance with Regulations

Noncompliance with applicable regulations can lead to consent orders, fines, possible shutdown of the utility operations, and lower credit ratings. A risk to a utility is its ability to meet the demands of the regulatory environment aimed at protecting public health and the environment, especially given the fact that the capital requirements for regulatory compliance are substantial.

During the rating process, rating agencies will consider the number of incidences of maximum contaminant level (MCL) violations. The nature of regulatory problems, consent orders the system may be under, and management action to address the problems are examined, as these items directly impact the operational and capital needs of the system.

Ensure that the utility's nonrevenue water percentage is within industry norms. Nonrevenue water is the difference between the finished water produced and the usage billed to the utility's customers. In accordance with accepted utility industry standards, nonrevenue water should typically be 15 percent or less for a well-managed utility. If the system has a significant number of line breaks and leakage, there may be higher operation and maintenance costs. The utility management should be prepared to discuss the nonrevenue water percentage with the credit rating agencies. If it is high, management should be able to provide reasons (e.g., leaks, need to flush lines to maintain the chlorine residual, construction water use, slow-registering meters that understate use, fire hydrant testing) and to outline any future plans to reduce the nonrevenue water.

Limit nonutility-related transfers to the general fund. Credit rating agencies consider that a best management practice is for the utility to have limited exposure to the financial operations of the general government so that system revenues can be used to operate and improve the utility. General governments should routinely complete studies of central services indirect costs to identify and justify the cost of providing support services to the utility enterprise fund, and the utility should review such studies to ensure the costs are reflective of the services provided. For transfers to the general fund, policies that limit the scope and growth of the transfers are favorable. However, any indirect cost reimbursements to the general fund should be considered a component of the utility's full cost of service.

Implement Regular Reporting and Monitoring

A utility's information systems are critical to effective management of the utility. They should have the ability to produce accurate and timely reports to utility managers to enable the managers to make informed decisions on utility operations. The utility's billing system should be established to identify any outliers in the customer data or statistics. Ideally, the billing system should also have the ability to produce bill frequency reports, which contain a summary of the number of bills rendered at various consumption levels during a given historical period, to enable the utility to estimate how rate structure changes could affect user rate revenue.

Track Receipts and Enforce Penalties Against Late Payers

The utility should have collection policies for delinquent accounts as well as policies for terminating service and placing liens on property for such delinquent accounts. The utility should also have an established practice of acting on such policies. If the utility's accounts receivable or uncollectible write-offs are significantly high in relation to the utility's billing cycle, the rating agencies may have credit concerns.

Incorporate Best Practices into Bond Issue Documents

The bond resolution, ordinance or indenture authorizing the issuance of bonds is a binding agreement between the regulators and the bondholders. As such, if certain best financial practices are required through this agreement, there is less implied financial risk.

Require renewal and replacement funding. Deferred maintenance increases the risk of system failure. Therefore, a utility that routinely funds renewal and replacement needs will be looked upon more favorably by rating agencies. Bond resolutions, ordinances, or indentures can require annual renewal and replacement funding. Required deposits equal to a percentage of the previous year's gross revenue are common. Ideally, the deposits should be linked to a comprehensive asset management plan that identifies all plant renewal and replacement needs for the utility system.

Have a closed-loop system. The utility system can be made a closed-loop system through the flow-of-funds provisions of the bond resolution, ordinance, or indenture. In a closed-loop system, all revenues generated through system operations generally must remain within the system and can only be used for lawful purposes of the system (i.e., cannot be used for nonutility-related purposes). Rating agencies prefer closed-loop systems since monies remaining in the system can be utilized to improve the system and pay for debt service and other financial obligations.

Plan for Contingencies

Prioritized spending and contingency plans are important risk management tools for utilities, as a utility must be able to adjust to economic, environmental, and regulatory changes that can affect revenues and expenditures. It is appropriate for adequate contingency allowance to be included in utility budgets.

If the utility applies for a rating through a bond issuance, rating agencies will stress test the financial projections. Stress testing involves adjusting one or more variables in the financial plan to assess the financial performance of the utility under a stress condition, such as a significantly reduced growth forecast or a higher capital investment need. Utilities should perform stress tests internally as part of the financial planning process. The standard stress tests that Fitch Ratings uses during the rating process to evaluate potential performance during periods of intense operating weakness (e.g., prolonged drought, economic softness, and/or a decline in customer usage) include the following scenarios:

- A precipitous drop in revenues that continues throughout the financial forecast period without corresponding decreases in expenses or increases in rates.
- A similar initial decline in revenues, followed by an immediate increase in user charges the following year to an affordable level that would lead to an increase in revenues during the remaining years of the financial forecast period.
- Break-even scenario: The level of revenue declines that could be sustained without cuts in expenditures.

Other stress tests that Fitch Ratings might use include:

- Calculating variable-rate debt service at a higher interest rate in future years than that assumed by the utility, if more than 20 percent of the utility's debt profile consists of variable-rate debt.
- Assuming a higher level of operating expenses than that assumed by the utility to account for potential budgetary pressures, such as accelerating chemical, electricity, or purchased water costs (Fitch 2012).

Recognize Additional Bonds Test

If a utility is attempting to secure additional funds through the issuance of revenue bonds, the utility is typically required to demonstrate that historical net revenue—often for a selected 12-month period within the past 24 months—can cover the maximum annual debt service (MADS) on both outstanding and new bonds. The MADS coverage test promotes interperiod equity and helps to prevent utilities from shifting an excessive amount of responsibility for debt payments to future ratepayers.

Additional bonds tests usually allow the utility to adjust the historical period revenue to reflect any changes in operations resulting from the new issue, as well as any rate increases that have been adopted prior to the bond issuance. If all the rate increases needed during the forecast period are not adopted as recommended, then the governing bodies must at least adopt sufficient adjustments to meet the additional bonds test requirements. If the additional bonds test coverage is not met, the utility cannot issue the bonds.

Be Prepared to Provide Facility Tours

When a utility applies for a bond rating, rating agencies will interview key personnel of the financing team. These interviews can be conducted through teleconference calls or in person at either the rating agency's office or at the utility. Scheduling an onsite visit has obvious advantages, as rating analysts can physically tour the utility's facilities. If the facilities appear well-managed, the analysts may have a more favorable impression of the utility's creditworthiness.

SUMMARY

Utility managers should have a fundamental understanding of how third parties evaluate the utility business, and should try to adopt and implement the management practices of highly rated or financially strong utilities. Many of these practices require a significant amount of time to implement and, as such, utility managers must plan ahead and adopt a long-term perspective when making operational and financial decisions for the utility. By following the guidelines of this chapter, utility managers can position the utility so it is looked on favorably by rating agencies and lending institutions and can secure lower interest rates when financing its capital needs. Moreover, by implementing financially sound practices, the utility can promote the viability of the enterprise and help keep user rates affordable for the long-term benefit of its customers.

REFERENCES

American Society of Civil Engineers (ASCE). 2013. Report Card on America's Infrastructure. www.infrastructurereportcard.org/fact-sheet/drinking-water. Accessed March 21, 2013.

American Water Works Association (AWWA). 2012. *Buried No Longer: Confronting America's Water Infrastructure Challenge*. Denver, Colo.: AWWA. www.awwa.org/portals/0/files/legreg/documents/buriednolonger.pdf. Accessed May 19, 2014.

Fitch Ratings. 2007. *Water and Sewer Revenue Bond Rating Guidelines*. New York: Fitch Ratings.

Fitch Ratings. 2010. *Revenue-Supported Rating Criteria*. New York: Fitch Ratings.

Fitch Ratings. 2012. *U.S. Water and Sewer Revenue Bond Rating Guidelines*. New York: Fitch Ratings.

Fitch Ratings. 2013. *2013 Water and Sewer Medians*. New York: Fitch Ratings.

Internal Revenue Service. Internal Revenue Code 148. Arbitrage, Reasonable Expectations—Artifice or Device, Investment Property, Required Rebate to the U.S., Small Government Unit Exception, Penalty in Lieu of Rebate. (26 USC § 148 – Arbitrage). www.irs.gov/Tax-Exempt-Bonds/Section-148-Arbitrage-Reasonable-Expectations-Artifice-or-Device,-Investment-Property-Required-Rebate-to-the-US-Small-Government-Unit. Accessed May 16, 2014.

Larkin, R. P. 2005. Wall Street Perspectives on Management that Matters, *Management Innovation in U.S. Public Water and Wastewater Systems*, Hoboken, N.J.: John Wiley & Sons.

Moody's Investors Service. 1992. *Moody's on Revenue Bonds: The Fundamentals of Revenue Bond Credit Analysis*. Boston, Mass.: Moody's Investors Service.

Moody's Investors Service. 1999. *Rating Methodology: Analytical Framework for Water and Sewer System Ratings*. Boston, Mass.: Moody's Investors Service.

Standard & Poor's Financial Services. 2002. *Public Finance Criteria Book*. New York: Standard & Poor's Financial Services.

Standard & Poor's Financial Services. 2009. *U.S. Public Finance: Key Water and Sewer Utility Credit Ratio Ranges*. New York: Standard & Poor's Financial Services.

Sugden, John, and Robin Prunty. Standard & Poor's. The Top 10 Management Characteristics of Highly Rated U.S. Public Finance Issuers. *Government Finance Review*. August

2012, pp. 44–45. www.gfoa.org/sites/default/files/GFR_AUG_12_44.pdf. Accessed May 19, 2014.

US Environmental Protection Agency (USEPA). 1997. Combined Sewer Flows—Guidance for Financial Capability Assessment and Schedule Development. EPA 832-B-97-004.

Chapter 5

Going to Market: The Process

This chapter focuses on the issuance of bonds. For the vast majority of government-owned utilities, the most common type of debt is the tax-exempt bond.

After examining external funding alternatives, bond issuance is a potential next step in the implementation of a capital improvement program (CIP) and an overall strategic plan. Financing goals and objectives should be set within parameters that enhance a utility's operational effectiveness. At a minimum, goals relating to financing costs, flexibility, bond term, and integrity should be established at the start of the financing process. These goals might include:

- Front-end financing costs need to be included in the budget. Annual debt service (principal and interest) may be included within an existing rate or financial planning study.
- For maximum flexibility, bond covenants should not restrict the utility in financing future capital improvements, setting rates, or operating the system according to the CIP and the overall strategic plan.
- The bond term should be set so the final maturity of the bonds do not exceed the estimated useful life of the project(s).
- The financing process should be carried out with integrity, in a manner that protects the public trust in the utility.

Although the utility may have the assistance of various outside professionals throughout the financing process, the utility's governing body remains legally responsible for the actions taken and information provided to investors. Utility management needs to control the process and secure the most cost-effective results. Staff members who understand the overall financing process and can effectively manage the finance team, including outside professionals, will be more effective agents of the governing authority. The roles and responsibilities of each team member are discussed in chapter 6.

OBTAINING AUTHORITY

Before a government-owned utility can issue long-term debt, it must obtain authorization from its governing body—city council or town governing body, utility commission, or regional body. To obtain approval, the utility should present

- the multiyear CIP,
- the proposed CIP funding plan, and
- the long-range financing plan covering payment of debt and annual utility operations.

Often a financial plan or rate study is used to demonstrate to the governing body how effectively existing or new rates will support outstanding and additional bonds. In some cases, a cost-of-service study may be necessary to show financial impacts on customers served. Before the governing body approves the bond issuance and rate study, a public hearing may be required. At the hearing, the governing body would provide ratepayers and other stakeholders the opportunity to speak for or against the utility's capital program and proposed rates. If debt is to be secured by the taxing power of the state or local community, some utilities must obtain approval by voter referendum of issuance on a local or statewide ballot.

Generally, the engagement of professionals, the authorizing bond resolution of the borrower, and the bond purchase agreement (BPA), the contractual agreement between the borrower and the initial purchaser, will need approval by the utility's governing body. The BPA is often adopted as part of an award resolution incorporating details of the financing terms.

In most (but not all) states, government-owned utilities do not have to file with the state's public utility or service commission. Investor-owned utilities must file with and obtain approval of the state agency and may be required to obtain Securities and Exchange Commission (SEC) approval before issuing the long-term debt.

ASSEMBLING THE FINANCE TEAM

After the governing body has approved the CIP (or refunding) and has instructed staff to assemble a finance team to assist in implementing the financing plan, the internal resources of the issuer should be evaluated, as discussed in chapter 3. The staff should also evaluate the status of any outstanding commitments or contracts with outside professionals. It is not uncommon for issuers to engage a bond counsel, financial advisor, or both for a period of time rather than for a specific issue. For the purpose of the following discussion, it is assumed that the issuer has no existing relationship with a bond counsel, financial advisor, or underwriter. However, it is also assumed that the project principals—consulting engineer and feasibility or rate consultant—have established working relationships with the utility.

Because the financial advisor serves the issuer in a fiduciary capacity, it may be prudent to select this professional first for guidance in choosing the bond counsel and underwriter. The selection of these professionals should be based on the results of an open, merit-based request for proposals, as well as a ranking system that includes the following criteria:

- Firm's related experience
- Specific experience of the assigned professionals with similar utility financings
- Available resources of the firm to serve the issuer (e.g., personel and their existing workload)
- Proximity of the firm to the issuer to be available to staff and governing bodies
- Level of compensation and structure (e.g., hourly, flat fee, is it contingent on completing a borrowing)
- Knowledge of the issuer's existing debt structure, future needs, and innovative ideas for the issuer

The selection process may include a short list of two to three firms based on written responses to the request for proposals. After conducting formal interviews of the short-listed firms, the utility should compile a final ranking for approval by the governing body.

The structure of compensation may be more important than the level of compensation in selecting a financial advisor and bond counsel. Expertise and level of service can differ greatly among firms applying for these important positions on the finance team. The utility should compare the structure and level of compensation with those of similar transactions by other issuers.

Before selecting an underwriter, the finance team should determine the type of sale. An underwriter is needed only if the issuer determines to sell the bonds by means of a negotiated sale. Negotiated sales are addressed later in this chapter.

TIMELINE

Assembling the finance team could take from 60 days to 6 months. This team, once assembled, should compile a timetable through bond closing that lists each task to be accomplished, the responsible team members, and a corresponding due date for each task. A timetable, although it should be flexible, will help the individual finance team members focus on their responsibilities and keep the process moving forward. An example of the process and time frame of issuing bonds is illustrated in Table 5-1.

Before working with rating agencies, bond insurers, and potential investors, the finance team may be required to have the following documents in substantially final form. (However, not all utilities are required to obtain an independent engineer's report or a financial feasibility report to issue bonds.)

- bond resolution
- engineer's report
- financial feasibility report
- preliminary official statement (POS)

The bond resolution, engineer's report, and financial feasibility report may be drafted concurrently. If a rate study has recently been completed, preparation of these documents may take from 3 to 12 weeks to update the findings based on the most recent utility experience, depending on a variety of factors, including the size of the utility and proposed capital funding requirements. However, if the rate study and financial plan are outdated, three to four times this length of preparation may be necessary.

Table 5-1 Sample timeline for bond issuance

Bond Issue Insurance Tax			Weeks and Responsible Parties											
No.	Description	Days*	1	2	3	4	5	6	7	8	9	10	11	
1.	Hold first meeting of finance team to discuss timetable and other matters	1		All										
2.	Distribute first draft of engineer's report, financial feasibility report, and bond resolution	20				ENG, FFC, BC								
3.	Hold conference call or meeting for document review	3				All								
4.	Distribute second draft of engineer's report, financial feasibility report, and bond resolution, and first draft of preliminary official statement	6					ENG, FFC, BC, DC, AUD							
5.	Conduct second document review	3					All							
6.	Send packages to the bond insurers and rating agencies	7						FA						
7.	Make site visits, road-trip presentations, or both, to bond insurers and rating agencies	10						All						
8.	Receive ratings and insurance commitments	7							FA					
9.	Distribute final preliminary official statement	3							DC					
10.	Print and mail preliminary official statement	2								DC				
11.	Market bonds	12									UND			
12.	Price bonds	1										UND		
13.	Award bonds	1											All	
14.	Print and mail final official statement (OS)	2												DC
15.	Conduct closing	12											All	

*Days to complete task

Shaded area indicates the task schedule: AUD—Auditor, BC—Bond counsel, DC—Disclosure counsel, ENG—Engineer, FA—Financial Advisor, FFC—Financial feasibility consultant, UND—Underwriter

Because the bond resolution, engineer's report, and financial feasibility report provide the basis for the POS, a working draft of the POS is usually the last document available for review. If the transaction is on a fast track, the documents may be sent to the rating agencies and bond insurers with some missing data, but they should contain no material inaccuracies. The use of bond insurers has decreased significantly since 2007. The financial advisor can assist the issuer in determining whether using a bond insurer is economically advantageous.

After a package of documents is sent to the rating agencies, a formal response may take approximately 10 to 21 days. Utility management should offer to meet with these entities to make a formal presentation of the package.

After the ratings have been secured, the POS can be printed and mailed to investors. This action commences the marketing period of approximately 3 to 14 calendar days for the bond sale. Bond closing usually follows the sale by 7 to 14 days.

When soliciting competitive bids for new bond issues, many utilities use the Internet to expedite the transfer of official documents, minimize costs, and reach as many prospective bidders as possible.

On the date of the bond sale, the terms, interest rate, price, and amount of the transaction are agreed to by the issuer and the underwriter. The actual exchange of bonds for dollars takes place at the closing.

Many states have a validation process during which the issuer may file a validation complaint with the state court. This filing provides notice and a forum for interested parties to challenge the bond resolution and the issuance of bonds. Basically, this process provides the issuer protection from legal challenges subsequent to the issuance of bonds. Normally, the procedure takes 60 to 90 days and can run parallel to other tasks, except closing, once the resolution is available. The issue of validation should be discussed with the bond issuer's own counsel or bond counsel, or both.

PREPARING THE OFFICIAL STATEMENT

The official statement (OS) is the document that presents a utility to the underwriter and then to the ultimate investor in the bonds. It provides the following information:

- Description of the bonds
- Description of the project being financed
- How the bond proceeds will be used
- A review of the management and organization
  - Organization chart with key personnel data
  - Financial, operating, legal, and tax information
- Description of the system and operations
  - Customers served
  - Size of the system
  - Regulatory compliance record
  - Rates and billings
- Financial results and projections

When prepared, the engineer's report, the bond resolution, and the financial feasibility report are the primary sources of information contained in the OS. The SEC has promulgated certain regulations requiring the issuer to certify that the OS, as of the date of its publication, does not contain any untrue or misleading remarks or misrepresentations of material facts and does not fail to state any material fact that may result in the other statements contained therein being misleading.

When bonds are sold through a competitive sale, the OS is put together by the issuer, financial advisor, consulting engineer, financial feasibility consultant, bond counsel, and the utility's independent auditor or accountant. In both competitive and negotiated sales, although compiled with the assistance of others, the issuer has the ultimate legal responsibility for the accuracy of information in the OS. In a negotiated sale, the underwriter with particular assistance of underwriter's counsel will perform a due diligence review of the issuer and the OS.

The OS is prepared in two phases. Most of the work is done in the first stage, resulting in a POS. This POS, usually printed, is distributed to rating agencies, prospective buyers, and underwriters who, in the case of a competitive sale, are the prospective bidders on the bonds. Rating agencies typically will have the most questions about the POS for the issuer to address. Once the bonds are awarded to an underwriter, the statement is updated with prices received from the sale, including interest rates and yields for each serial and term bond, cost of issuance of the bonds (including cost for the underwriting), debt service information, and any other material information that was not a part of the preliminary statement. The final OS is then printed. The issuer bears the cost of printing and electronically distributing both statements.

METHOD OF SALE

Issuers have three methods available for issuing bonds:

1. Private placement
2. Publicly offered
3. Competitive sale

An issuer should review each method in light of financing costs (including interest), level of service, and political considerations.

Private Placement

In a private placement, the issuer sells bonds directly to a small number of sophisticated investors who have the ability to perform their own assessments of the issuer's credit. If the issuer's credit is deemed less than investment grade by a rating agency, the issuer might find it beneficial under certain circumstances to consider a private placement.

Because bonds sold through private placement are often not liquid to the investor (i.e., the investor is restricted from selling them or is often unable to sell them), the transaction could result in a higher effective interest cost. Historically, only a small percentage of the total dollar amount of tax-exempt bonds each year are privately placed. This financing method, however, has become more prevalent in recent years.

Publicly Offered

Bonds sold on a competitive or negotiated basis are publicly offered to a variety of potential investors. To the issuer, a bond issue appears to be a large debt obligation with a specific interest rate. However, publicly offered bonds are traditionally sold in $5,000 denominations with different interest rates for each maturity date to appeal to various investor groups. Therefore, each bond issue is typically purchased by numerous investors. The Federal Reserve System reported that approximately $3.7 trillion of tax-exempt bonds were outstanding in the United States as of Sept. 30, 2012. The percentages of ownership for different investor groups summarized in Figure 5-1 can vary over time, depending on changes in market conditions affecting other investment options.

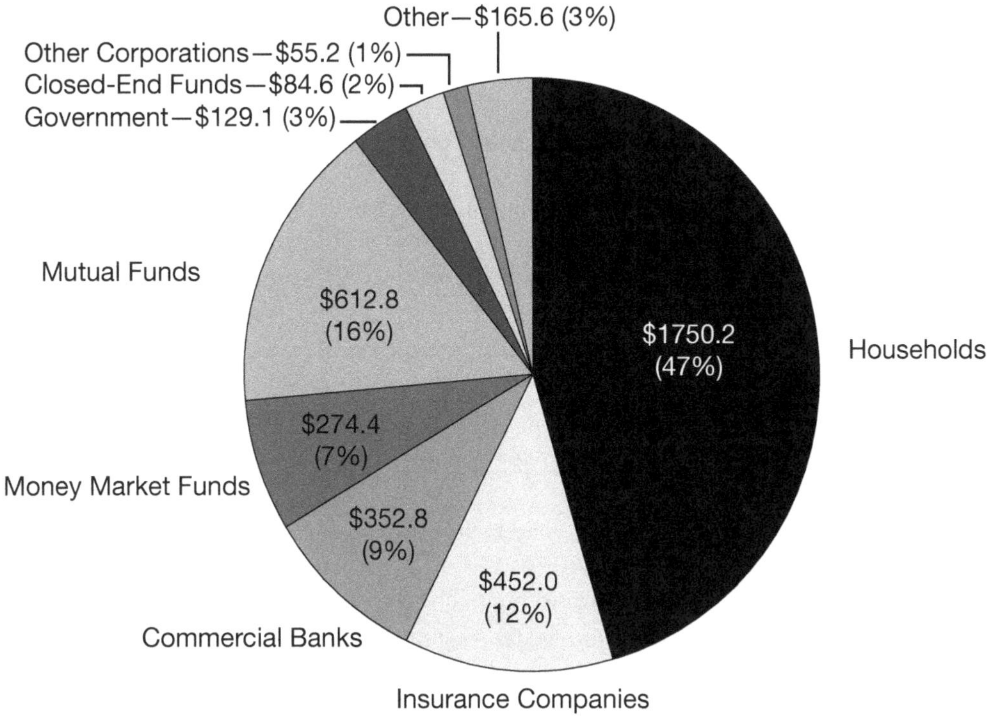

Note: All amounts in billion $US.
Source: 3rd quarter 2012; from Federal Reserve System Dec. 6, 2012 report.
Figure 5-1 Investor groups' holdings of municipal securities

Negotiated Sale

A negotiated sale involves selecting the underwriting firm or firms in advance of the sale. In a negotiated sale, the underwriters and other professionals assist the issuer in structuring the debt obligation. The issuer negotiates the price for the issue with the underwriters at the time of sale. Before the sale of bonds is negotiated, the justification for choosing the negotiated form of sale over the competitive sale should be developed. A negotiated issue can be justified based on reasons that include:

- A volatile market
- An unusual bond structure
- A refunding of bonds
- An issuer that is in the market infrequently

A financial advisor who has a feel for the market can be helpful in identifying facts that could provide justification for a negotiated sale. With a negotiated sale, the underwriter often assists in preparing the OS and other bond documents used to market the bonds. A disadvantage of a negotiated sale is that the process of establishing the price and interest rates for the bonds is not as straightforward as in a competitive sale. An issuer must evaluate the price and interest rates offered by comparing the issue against other equivalent issues offered just before the sale of bonds. A financial advisor is often used to help the issuer evaluate the terms and conditions of the underwriter's offer. Table 5-2 provides a comparison of conditions favoring either negotiated or competitive sales.

Table 5-2 Conditions favoring a method of sale

Debt Structure	Competitive Sale	Negotiated Sale
Security Characteristics		
Pledged revenues	General obligation or strong system revenue	Project-supported revenues
Security structure (for revenue bonds)	Conventional resolution and cash flow; rate covenant and coverage	Unusual or weak covenants; subordinated debt
Debt instrument	Traditional serial and term, full-coupon bonds	Innovative structuring; derivative products; targeted specific investors
Credit Quality		
Rating level	A or better	Below Single A
Rating outlook	Stable, A, or better	Weak, but improving; or under stress
Issuer Characteristics		
Type of organization	Broad-based, general-purpose borrower	Special-purpose, independent authority
Frequency of issuance	Regular borrower in public market	New or infrequent issuer
Market awareness	Active, informed issuer staff	Issuer with significant financial, legal, or other problems
Investor comfort	Well-known, stable issuer	Issuer with significant financial, legal, or other problems
Market Conditions		
Current trend	Stable, predictable, strong market	Volatile or declining market
Supply and demand	Light issue supply; good investor demand	Oversold market; heavy issue supply
Policy Considerations		
Participation in sale of bonds	Desire to have broad market participation for sale of bonds	Desire to direct business to disadvantaged businesses or to local or regional firms
Stimulation of investor interest	Desire to have broad market participation for purchase of bonds	Desire to direct business to local or regional investors

Competitive Sale

The competitive form of sale is the easiest and most straightforward method to sell a municipal debt obligation. It differs from the negotiated sale in that the issuer does not select the underwriting team but rather publishes (1) a notice of sale that specifies, among other things, how bids are evaluated (usually on the basis of true interest cost), and (2) a bid form that is delivered or made available with the POS to all prospective underwriters. This process is completed through electronic services that post the offering documents and bid form. The financial advisor and bond counsel review the third-party system to ensure any parameters of the bonds are correctly reflected. Underwriters submit their bids through an electronic service and the results are made available to the issuer and financing team.

The underwriters form underwriting teams, called *syndicates,* and submit bids on the bid form at the time and place specified in the notice of sale. The issuer opens the bids at the prescribed time and, after compiling the results, awards the sale to the underwriting team that submits the lowest interest cost as specified in the notice of sale.

Using this method, an issuer is assured a competitive or market rate on the bonds. In a competitive sale, the bidder has the option of obtaining credit enhancement by purchasing bond insurance. The cost of this insurance can be borne by the underwriter or the issuer as specified in the notice of sale. The underwriting team will use bond insurance if it believes insurance will produce a lower bid. Also, some investors are interested in purchasing only bonds of a high credit rating, and bond insurance might provide such a rating. As stated earlier, the availability of bond insurance may be limited and should be reviewed by the issuer and its advisors.

UNDERWRITER CONSIDERATIONS

Underwriters enter the bond process in two ways. In a competitive sale, they bid on the bonds at the time of sale. The bid that produces the lowest net or true interest cost, as specified in the notice of sale, determines who the underwriter will be for the issue. In a negotiated transaction, the issuer actively selects the underwriter for the issue.

Depending on the size of the issue—small, medium, or large—the issuer decides along with the financial advisor, if any, or the senior underwriter, whether the issue needs co-managers. A senior manager negotiates the terms and conditions for the bonds directly with the issuer and assists in establishing the best time to sell the issue. The senior manager keeps the books and allocates the bonds to the various managers. The underwriter also has due diligence obligations promulgated by regulators. The underwriter, with the assistance of underwriter's counsel, will conduct a due diligence review of the issuer and the information contained in the OS. This review may be conducted through a conference call or an onsite visit at the offices of the issuer. Often the issuer will need staff heads from multiple disciplines to provide answers in the due diligence process. The issuer, along with the financial advisor, oversees this activity.

The use of co-managers helps in gaining access to various segments of the market that might not be thoroughly covered by the senior manager. Some underwriting firms have the sales strength to access institutional markets, whereas others have access to retail markets. The markets that the issuer is trying to reach often determine the choice of co-managers. In some cases a selling group of additional underwriting firms will be established by the financing team to provide an additional level of access to the market. Selling group members usually assume none of the underwriting risk and do not cost the issuer any additional money.

SALE AND CLOSING

The sale and pricing constitute the most important event in the financing process. This is when interest rates and other terms of the transaction are determined. The underwriter will offer rates and terms that are supported by offers from potential investors. Because the investors hold offers open for only a short period of time, the issuer's staff, with the financial advisor's assistance, should be prepared to give a verbal award within minutes after bids are taken in the case of a competitive sale or on conclusion of discussions in a negotiated transaction. Formal approval of the governing body generally takes place within 24 hours of the verbal award unless the governing body has previously delegated that authority by formal approval.

A period of time is predetermined during which all the documents of sale—including legal and tax documents, plus those required by the governing resolution—must be completed. On the prearranged date, these documents are executed and the sale is finalized. To avoid last minute problems, most documents are executed in a preclosing held before the actual date of closing. On the date of closing, the underwriter usually transfers the net bond proceeds to the appropriate bank or trust company designated by the issuer, and the issuer simultaneously delivers the bonds to the underwriter.

POSTSALE CONSIDERATIONS

After the sale is completed, the finance team should evaluate the success of the sale in relation to the goals and objectives set in the initial stages of financing. For example, the team should answer the following questions:

- Do the true interest costs and debt amortization conform to the criteria identified in the reports used to inform the governing body and public?
- Did the issuer retain the flexibility to easily finance future utility capital improvements?
- Did financing costs stay within estimates and bids?
- Were credit ratings of the issuer retained, downgraded, or enhanced?
- Did the process enhance relations between the issuer and investors?
- Did each team member perform as promised?
- Did the team work well together?
- How can the process be improved?

By documenting the answers to these and other questions, an issuer can enhance the success of future financings. Many issuers also find it helpful to have the finance team continue to meet on a periodic basis to improve communication and to better manage the capital improvement funding process. This communication also improves the issuer's ability to monitor the outstanding debt for possible refunding opportunities.

The public offering carried out by the issuer is not the last time the bonds are sold. Initial investors frequently sell bonds to other investors in the secondary market. To protect all investors, the SEC adopted continuing disclosure requirements that were effective July 1995. These requirements call for issuers to update material information contained in the OS on an annual basis and to provide notice of any of the following 11 material events. Bond counsel or the financial advisor should know whether these apply.

1. Principal and interest payments delinquencies
2. Nonpayment-related defaults
3. Unscheduled draws on debt service reserves reflecting financial difficulties
4. Unscheduled draws on credit enhancements reflecting financial difficulties
5. Substitution of credit or liquidity providers, or the failure of these providers to perform
6. Adverse tax opinions or events affecting the tax-exempt status of the security
7. Modifications to the rights of security holders
8. Bond calls
9. Defeasances
10. Release, substitution, or sale of property securing repayment of the securities
11. Rating changes

The required information has to be filed with the Municipal Securities Rulemaking Board's Electronic Municipal Market Access (EMMA) to ensure its availability to all potential investors. EMMA is an Internet repository where primary and secondary disclosures are made available to the general public. The full finance team should identify compliance issues pertaining to SEC's continuing disclosure requirements early in the process, with the objective of providing ongoing adequate disclosure while not unduly burdening the issuer subsequent to the bond closing.

After the transaction has been completed, bond counsel usually provides each finance team member a transcript containing copies of the bond documents associated with the transaction. This serves as an easy reference tool throughout the life of the bond issue.

SUMMARY

Bond issuance is one step in a CIP's implementation. It involves obtaining the authority to issue a bond, assembling a finance team, and compiling a timeline to schedule the tasks. The official statement is prepared by the finance team for a negotiated sale or competitive sale. A limited offering document may be put together depending upon the type of private placement. During the sale, interest rates and other terms of the transaction are determined, making the sale the most important event of the financing process. After the sale is completed and the documents executed, the finance team should evaluate the sale in relation to the issuer's goals and objectives established at the beginning of the process.

Chapter 6

Going to Market: The Players

Many participants can be involved in the financing of a utility, so this chapter is dedicated to providing a broad explanation of the roles these individuals can play. The descriptions that follow identify the participants in the process and offer a synopsis of the responsibilities of each participant as a member of the finance team, the organizational role for each participant, and other pertinent information.

FINANCE TEAM

The financing staff of the issuing entity should have the responsibility for the bond-financing program. Regardless of how many outside consultants are retained, the issuer is responsible for all aspects of the program. The lead finance staff is responsible for all financial information, rates and revenues data, and all statistical data (e.g., population growth and local economic issues). Because most of the official statement (OS) is technical in nature, including the names, titles, roles, and contact information of issuer, identifying the personnel responsible for operating and engineering areas is critical.

After the expertise for the financing team is determined, outside consultants should be selected. Potential team members and their roles are listed in Table 6-1.

Typically, successful financing is the result of cooperative efforts of the issuer's professional staff working in conjunction with consultants representing the legal, financial, and engineering disciplines. This group forms the finance team—the individuals charged with the responsibility of developing and implementing bond financing programs.

The composition of the finance team is based on the financing requirements. A public bond offering, for example, is the most complicated financing process, particularly when compared to a private placement. It requires the collective expertise of professionals from

Table 6-1 Potential finance team members

Team Member	Role Description	Reports to/ Represents
Issuer (utility)	Determines need for bonding; selects financing team; responsible for all financial information, and typically includes utility finance director, engineers, and operations leads. Finance director may be required to develop coverage test(s).	Governing body, Customers
Legal Principals		
Bond counsel	Provides expertise in securities law; provides an opinion regarding the tax status of bonds	Issuer
Issuer's counsel	Reviews local legal matters for inclusion in the OS	Issuer
Disclosure counsel	Conducts due diligence; may draft the OS; provides guidance for ongoing disclosure responsibilities	Issuer
Underwriter's counsel	Reviews all documents; may prepare OS and other documents of concern to the underwriter	Underwriter
Financial Principals		
Financial advisor	Provides financial analysis assistance with the issue	Issuer
Underwriter	Purchases the bonds and resells to investors	Issuer
Commercial bank	May offer tax-exempt loans or credit facilities to issuers	Issuer
Project Principals		
Financial feasibility consultant	Develops a financial feasibility analysis of the utility and/or project	Issuer
Consulting engineer	Provides analysis on operations and maintenance (O&M) of the system and new projects	Issuer
Credit Evaluation Principals		
Rating agency	Rates creditworthiness of entity for investors	
Credit enhancer	Insures the bonds or provides an LOC* to back the bonds	Issuer
Trustee	Holds funds and is responsible for protecting investors' rights	Bondholders
Auditor/certified public accountant (CPA)	Provides most recent audited financial statements and may provide escrow verification	Issuer
Other Participants		
Financial printer	Prints OS, notice of sale, or other requested items	Issuer
Verification agent	Verifies the accuracy of the bond yield and cash flows related to advance refunding	Issuer
Registrar	Maintains records of bond ownership	Issuer
Paying agent	Disperses principal and interest payments of bondholders	Issuer

*Letter of credit

many different areas. In the form of a public offering, it necessitates extensive legal documentation, preparation of disclosure materials, selection of an underwriter, and meetings with the rating agencies and bond insurers. Private placements or bank loans, on the other hand, are based on proprietary credit evaluation and can be completed on standardized loan documents. Both types of financing are important and require the organization of a finance team to ensure a successful closing.

The composition of the finance team is based on the financing requirements. It is incumbent on the issuer and consultants to mobilize the resources necessary to organize an efficient finance team.

LEGAL COUNSEL

Legal counsel will fulfill a number of diverse roles in the course of any financing program. Documentation and legal opinions are required on myriad issues ranging from the legal organization of the borrowing entity to the legality of an investor in another state purchasing bonds. Generally, two to four attorneys are involved in a bond issue, depending on the complexity of the transaction.

Bond Counsel

A bond counsel is an attorney or firm of attorneys with specific expertise in securities and tax law. Bond counsel is engaged by the issuer and charged with the responsibility of issuing an approving opinion that the debt obligation is valid, legal, and binding. To do this, bond counsel prepares documentation authorizing the issuance of debt and determines any conditions with which the issuer must comply. Bond counsel may be involved with various documents, including a bond referendum, state legislation, and notices to bondholders, among others.

Perhaps the most important role for bond counsel is to provide an opinion regarding the tax status of bonds. The primary concern is whether interest paid on the borrowing is exempt from state taxes, federal taxes, or both. Federal law governs the issuance of tax-exempt bonds and provides specific direction concerning the types of projects eligible for tax-exempt financing, as well as refunding (refinancing) limitations. State laws generally focus on a narrower set of issues, such as the purpose of borrowing. It is incumbent on bond counsel to determine the extent of state and federal tax exemption, if any, for governmental borrowings.

Bond counsel's responsibilities continue after the transaction is completed. Because underlying documentation can be complex, bond counsel is frequently called on to provide interpretations and clarifications. This is an important resource when an issuer is monitoring compliance with covenants or conducting an audit.

Issuer's Counsel

The bond counsel works closely with the issuer's counsel, an attorney who may be an in-house staff member or may be retained on a contractual basis. In the course of a debt financing, the primary focus of the issuer's counsel is on administrative and procedural matters. Typical opinions provided by issuer's counsel relate to the legality of organization, the validity of meetings, the process by which bond issuance authorization is obtained, and proper adoption of required documentation, among others. Additionally, the issuer's counsel works closely with bond counsel to review documents and ensure that they are operable.

Disclosure Counsel

With additional regulations pertaining to disclosure by issuers of tax-exempt debt, a new role of legal counsel has evolved. Disclosure counsel is engaged by the issuer and is primarily responsible for conducting due diligence, drafting the OS, and providing guidance with respect to ongoing disclosure responsibilities. Often the task of drafting the OS is delegated to either bond counsel or underwriter's counsel. However, some of the larger issuers are moving toward the use of a disclosure counsel in response to continuing disclosure regulations recently promulgated by the Securities and Exchange Commission (SEC). Bond issuers have recognized that the OS is their disclosure document and want to have their legal counsel oversee its preparation. Notwithstanding this position, bond counsel or underwriter's counsel still may be delegated the responsibility of preparing the OS.

Underwriter's Counsel

Much has already been said about the duties of underwriter's counsel, and it is important to acknowledge that this type of attorney represents the interests of the underwriter. In this capacity, underwriter's counsel has certain document preparation responsibilities and reviews all documents and opinions prepared by other counsels involved in the transaction. In addition to the OS, underwriter's counsel prepares a purchase contract between the issuer and the underwriter, as well as various underwriting documents and securities filings that are of concern only to the underwriter. Underwriter's counsel is retained by the underwriter.

FINANCIAL PRINCIPALS

An issuer may call on several financial experts to provide assistance in the course of a financing program. Many of these experts possess the same level of financial skills but are distinguished by the role in which they serve the issuer. Financing professionals frequently serve as leaders for the finance team because they coordinate much of the work among attorneys and rating agencies.

Financial Advisor

A financial advisor may be involved in any type of debt-financing transaction, whether it consists of issuing bonds or undertaking a loan. The financial advisor serves the issuer in a fiduciary capacity and is required to be retained with a formal contract establishing specific responsibilities. This contract is significant because the issuer frequently acts on recommendations from the financial advisor, and there should be no questions about whose interests the financial advisor represents. Two different types of financial advisors counsel governmental entities on financial matters. One type is the traditional investment bank that provides underwriting and financial advisory services. The other is an independent firm that offers financial advisory services only, but does not underwrite issues.

Issuers can use a financial advisor's services on a broad range of projects. Within the context of this manual, however, the focus is on the development and implementation of efficient and cost-effective financing strategies. To provide these services, the financial advisor helps the issuer to evaluate the creditworthiness of alternative revenue scenarios and covenants. This might involve preliminary discussions with rating agencies or private placement purchasers to assess risk issues. This assessment helps the issuer understand the ramifications of various structural alternatives and supports the decision-making process.

The financial advisor is capable of assisting an issuer with negotiation and evaluation of various fees and costs, as well as interest rate proposals for notes, bonds, or loan products. An experienced financial advisor is knowledgeable about qualified vendors for such specialty areas as financial printing; trustee, paying agent, or registrar services; and escrow verification.

Compensation for the financial advisor can be determined on an hourly basis, through a transaction fee based on the amount of debt, on an annual retainer, or through a combination of these methods. The issuer should select the compensation method that best suits the needs of the issuer.

Accountant

For most bonds issued, an independent auditor or accountant provides a copy of the most recent audited financial statements, which are incorporated as an appendix to the OS. Summaries of this audited financial information are generally included in the body of the OS. The accountant may review the OS to ensure that all historical information is accurately presented, is referenced to the audited financial statements, and is evaluated in light of supporting schedules and footnotes.

The issuing entity's accountant or external auditors often (but not always) play a role in monitoring ongoing compliance with bond covenants as it relates to satisfaction of debt-service coverage requirements. Depending on the exact wording of the bond ordinance or covenants, a certified public accounting firm other than the utility external auditor may fulfill this responsibility.

When an escrow is established, as is the case when bonds are refunded, an accountant is engaged to perform the escrow verification. While the issuing entity's external auditing firm may provide this service, a separate firm is often engaged. The accountant may report to the finance director, or the financial advisor or underwriter may select the verification agent.

Underwriter

An underwriter has a different business relationship with the issuer. The underwriter is not contractually bound to act in a manner that serves the interest of the issuer. The issuer must look for ways to ensure that its own interests are protected while also maintaining a good working relationship with the underwriter. In addition, underwriters are required by regulators to obtain written acknowledgement of their role from an issuer as well as provide certain disclosures based on the complexity of the transaction that they are recommending.

An underwriter provides virtually the same analytical capabilities as a financial advisor. Beyond this similarity, the underwriter acts as an intermediary for an issuer to access capital markets. This market access provides the mechanism for the issuer to raise capital through the bond market by marketing debt securities (generally bonds) to institutional and retail investors. The underwriter negotiates the terms of sale with the issuer, uses its sales professionals to market the issuers' bonds, and purchases any unsold bond amounts.

Underwriters generally work on a contingency fee basis, gaining compensation only through the successful sale of bonds. They derive remuneration by purchasing bonds from the issuer at a discount and reoffering such obligations to the public at a higher price. There is rarely a need for an underwriter to invoice an issuer because the underwriter's discount should provide sufficient amounts for professional services, sales commissions to market the bonds, reimbursement of out-of-pocket expenses, and underwriting risk, if any.

Commercial Bank

A commercial bank may also be part of a finance team, depending on the nature of a borrowing. Bond issuance can be an expensive undertaking; for smaller borrowings, securing funds through a private placement or bank loan may be more cost-effective than a bond issue. When a commercial bank evaluates a loan application, it reviews many of the same factors considered by a rating agency or bond insurer (see the discussion of credit evaluation principals later in this chapter). However, one intangible factor for a local bank is knowledge of the local community, including familiarity with development plans.

Bank loans can be offered on a tax-exempt basis to traditional municipal issuers. Additionally, certain loans to governmental entities that issue less than $10 million per year can be designated as bank qualified. This is important because it allows a bank to write off a portion of the funding costs, making it possible for the bank to extend a lower interest rate.

Loans are not the only debt product commercial banks offer to municipal borrowers. While such banks traditionally underwrite certain types of debt (e.g., GO bonds), they are also becoming more active in providing credit facilities. These facilities range from a letter of credit (LOC) securing a bond to an LOC providing a source of funds on which to draw for construction expenditures.

PROJECT PRINCIPALS

Some utility financing may be considered to be project financing. This means that the cash flow produced by a given project is sufficient to meet operational and debt service obligations. To ensure that the project can and will continue to be self-supporting, the financial community requires independent professional studies pertaining to the technical and financial aspects of the project. These services may be obtained from the consultants described in this section.

Financial Feasibility Consultant

A financial feasibility consultant develops a multiyear financial plan for a utility. A feasibility study is intended to project a utility's cash flow based both on historical performance and on expected performance after the project improvements are put into service. The financial community has a vested interest in understanding a utility's ability to meet operational and debt obligations from pledged revenues.

A utility may choose among different types of professionals to prepare a financial feasibility report. For the most part, these consultants fit into one of two categories:

- Independent financial, economic, and accounting consultants
- Professional engineering firms with extensive backgrounds in preparing financial and engineering studies

The primary role of the financial feasibility consultant is to prepare projections that illustrate the impact that the proposed capital projects and associated debt service will have on the financial position of the utility. This is generally measured by calculating debt-service coverage ratios and projecting net cash or fund balance for a multiyear period, typically five years. The financial feasibility consultant may also provide the utility with other ongoing services required by the bond resolution, including annual determinations of net revenue. Through the issuance of a separate report or certificate, frequently referred to as an additional bonds or parity certificate, the financial feasibility consultant may be responsible for confirming compliance with bond covenants that test the issuing entity's ability to make future debt service payments. The certificate's resulting opinion is based on management's future plans and representations pertaining to estimates and assumptions.

An issuer can select either of these two consultant types as long as they possess the necessary credentials and experience. For the most part, there are no formal professional standards for the preparation of financial feasibility documents by general consulting and professional engineering firms. The financial feasibility study mostly forecasts revenue and costs into the future. It should describe the assumptions used in revenue projections (such as growth, rate increases, and interest rate used for earnings) and assumptions used for operations and maintenance costs (such as inflation, new facility cost, and variances).

However, when bond covenants require a certified public accountant's (CPA) seal on the financial feasibility study, the CPA must follow specific guidelines promulgated by the American Institute of Certified Public Accountants (AICPA). The underwriter, financial advisor, and any applicable state oversight body may be able to assist the utility in determining the appropriate party to provide the financial feasibility service. Depending on the nature of the service provided, the financial feasibility consultant may report through the utility's finance director.

A financial feasibility study is actually a compilation of information. The financial feasibility consultant reviews information to confirm that it is reasonable but requires most, if not all, of the data to be provided by the utility or others on the finance team. Generally, the consultant also assesses the reasonableness of the utility's projections of system growth, water sales, cost trends, and so on. This assessment pertains to operations and construction cost estimates, the service demand forecast, interest rate assumptions for bonds, and other costs. Depending on how the financial feasibility consultant is retained, the consultant may or may not be involved in developing projected rates or charges for the utility.

The financial feasibility report usually contains

- a detailed summary of existing rates,
- a description of any proposed rate changes and their impact in terms of dollars and percentages,
- the consideration of any alternative rate structures in light of customer concerns or utility objectives (e.g., affordability, water conservation), and
- A forecast of cash flow and debt-service coverage.

A utility's historical rates, current rates, and trend lines for historical rate increases are often compared with those of other utilities on a local, statewide, or national basis. The impact on and ability to use projected system development charges, a one-time revenue source, to satisfy debt-service coverage requirements may be evaluated.

Engineering Consultant

Most financial participants to a bond issue, including the underwriter, credit analyst, and investor, conduct extensive due diligence for each project. This normally begins with an objective assessment of the particular utility by a qualified engineering firm with recognized expertise in the water industry to confirm that the utility is effectively operating and maintaining the system. The utility is responsible for hiring the engineering consultant.

During the bond-issuance process, the engineering consultant issues a report that typically provides three assurances that become an integral part of the disclosure document and is generally included as an appendix to the OS. These assurances are:

- That the system is meeting the all federal, state, and local regulations by describing in detail existing and future regulations, how the system is complying, and steps being taken to comply in the future.
- That the utility is in good condition financially, that the project improvements are necessary and cost estimates are reasonable, that the construction timetable is appropriate, and that permits have been or can be obtained.

- That the utility can meet its projected financial obligations under bond covenants, shown by providing cash flow projections of future revenues and operating expenses.

Depending on the bond resolution, the consulting engineer may also be required to issue periodic reports or certificates regarding the condition and operations of the utility after the bonds are issued.

An engineering consultant's report ultimately evolves as a compilation of the many detailed reports that the utility's consulting engineer submits to the issuer concerning the development of the project. Local officials and utility managers have often used information reported at relatively early stages by a consulting engineer to make decisions on whether to proceed with capital improvement projects. A formal consulting engineer's report describes the plan selected for implementation and how such improvements will benefit the utility.

A formal report analyzes and draws conclusions relating to the utility's current and future operations. These results typically include an evaluation of the following aspects of the utility system:

- The age, condition, capacity, and adequacy of water supply sources and water treatment and distribution facilities
- Historical and projected water demand
- Water supply contracts (i.e., what percentage of the system's total supply a given contract represents and whether the contract term extends beyond the bond issue term)
- Future capital improvements and construction schedules and whether the construction plan is implementable and affordable
- How current and future water supply, treatment, and distribution facilities may impact the utility's service area

CREDIT EVALUATION PRINCIPALS

Rating Agencies

The complexity of the bond market renders traditional methods of credit assessment used by investors nearly impossible. By simplifying the credit rating and ranking of bonds, independent rating agencies have become an essential part of the process investors use in making bond purchase decisions.

Rating agencies have developed easily recognized grading systems that measure, on a single scale, bonds from municipal issuers ranging from large state agencies to smaller water districts. If a bond issue has high ratings from an independent rating agency, the investor's willingness to buy increases and the market for the bond broadens, reducing the cost associated with the bond issue. The resulting exposure to a wider segment of the bond market could reduce the bond interest the issuer pays.

As discussed in chapter 4, Standard & Poor's Financial Services, Moody's Investors Services, and Fitch Ratings are well-known independent rating agencies with widely used systems of ratings for municipal bonds. Although all independent rating agencies use many similar factors, a particular agency may put more weight on one factor in its analysis than another. The issuer should be aware of these differences when selecting an agency.

The rating agency's rating of a given bond is established as the bond is issued. Subsequently, the agency periodically reviews its rating until the bond is paid off. The rating expresses the quality of the bond and its credit risk. Key factors used to determine a bond rating are discussed in appendix C.

OTHER PRINCIPALS

Trustee

A trustee has several roles in the bonding process and subsequent activities. As a signatory to the bond resolution, the trustee, along with the issuer, has ongoing responsibilities with respect to the bonds. The primary role of the trustee is to act as a fiduciary of and to protect the financial interests of the bondholders. The trustee does not get involved with developing the bond structure but does substantially review final documentation to confirm that the document's provisions are administratively feasible. In addition, the trustee's legal counsel reviews and comments on the bond resolution and documents to make sure they meet requirements of the bond resolution, along with applicable legal and tax requirements.

At the day of closing, the trustee releases the bonds issued after verifying that the bond proceeds have been transferred to specified accounts. After the closing date, the trustee collects, holds, and disburses bond proceeds in accordance with the bond resolution. The trustee also collects, holds, and disburses debt funds.

The trustee's responsibilities extend to ensuring that the issuer meets all requirements of the bond resolution during the life of the bonds. A trustee is generally required for revenue bonds or certificates of participation (COP) but may not be required for a general obligation bond issue.

In addition, the *financial printer* publishes the offering document to potential investors on behalf of the issuer. The *registrar,* who is typically an employee of the trust company, maintains the official records of bond ownership, including resales on behalf of the issuer, and the *paying agent* disperses the principal and interest payments to bondholders on behalf of the issuer as well.

SUMMARY

This chapter has discussed the roles participants generally play in capital financing, and Table 6-1 provides a summary of potential team members, including general descriptions of their contributions. Regardless of the participants in the financing and the form that the finance team may take, it is important to remember that the issuer remains responsible for all aspects of the program.

This page intentionally blank.

Chapter 7

Special Considerations for Investor-Owned Utilities

The development of rates for investor-owned utilities (also called private utilities) differs from the rate-setting process for publicly-owned utilities (i.e., governmental water utilities) and has a direct effect on the funding of capital improvements by such utilities. Generally, publicly-owned utilities follow a *cash-based* approach for the development of the net revenue requirements to be funded from rates, which is consistent with the budget process and compliance requirements associated with the issuance of debt. Investor-owned utilities employ an *equity-based* ratemaking approach with the revenue requirements aligning more with financial reporting under generally accepted accounting principles. Table 7-1 shows the general differences in the development of the revenue requirements between publicly-owned and investor-owned utilities.

As shown in Table 7-1, the development of the net revenue requirements for investor-owned utilities does not generally have a direct or specific capital funding component identified in the rates for service. Rather, the capital component that is recovered in investor-owned utility rates is by the inclusion of depreciation expense and the return on equity as a component of the net revenue requirements. Accordingly, capital financing considerations for investor-owned utilities differ from those of publicly-owned utilities.

When identifying a capital-financing plan, investor-owned utilities must consider the availability of equity (both common and preferred stock) and the ability to issue debt to meet the capital needs. With respect to the payment of debt service on capital investment, including the interest cost on such debt, the source of funds is derived from the rate

Table 7-1 Determination of net revenue requirements

Publicly-owned Utilities—Cash-Based Approach	Investor-owned Utilities—Equity-Based Approach
Operations and maintenance expense	Operations and maintenance expense
+ Senior and subordinate debt payments	+ Net depreciation expense*
+ Capital funded from rates / Deposits to capital account†	+ Taxes other than income taxes
+ Payment-in-lieu-of-taxes to general fund	+ Income taxes
+ Other financial fund deposits‡	+ Rate of return on rate-base assets
– Other operating revenue	– Other operating revenue
– Investment income	= Net revenue requirements funded from rates
= Net revenue requirements funded from rates	

* Depreciation expense recognized in the investor-owned revenue requirements is generally net of plant (fixed assets) that has been contributed to the utility (either as a property contribution or a payment of a capacity-related fee) and plant that is considered as being non-used and useful for the provision of utility service (an adjustment frequently made to excess utility plant capacity or plant considered as being held for future use).

† Reflects ongoing capital funded from rates and deposits to capital accounts or funds such as a renewal, replacement, and improvement fund established as a condition of the issuance of bonds or by policy of local government/publicly-owned utility.

‡ Reflects deposits to other funds to maintain financial condition and rate compliance on bonds or set by financial policy of local government; may include deposits to operating reserves, rate stabilization funds, specific program funds (e.g., water conservation), and other funds established by the local government.

of return earned on the rate-base assets and the recovery of depreciation expense in the rates for water service. As such, the capital funding plan can be complicated by:

- The timing of when the utility will be allowed to earn a rate of return on the utility plant investment to be financed, which is linked to when the asset is completed and placed in service and can be included as a component of the rate base
- Lag time in implementation of rates due to the prolonged timeframe for the rate approval process by state public service commissions and other regulatory bodies
- The disallowance of operating expenses because the amount of constructed investment may not correspond to the current customer base (referred to as "non-used and useful plant")
- Increased costs that may result from changes in income and property taxes on utility investment

Investor-owned utilities, unlike publicly-owned utilities, have the ability to attract equity capital in a competitive marketplace. The ability to attract capital is directly related to the financial position of the utility, the dividend payments and capital appreciation shareholders expect to receive from their equity investments in the utility, the demographics of the utility service area, and the overall corporate structure of the utility. Many investor-owned utilities are a division of a larger corporation or utility holding company that can achieve certain economies of scale in operation and may result in a stronger financial credit with an enhanced ability to attract additional equity or debt financing in the marketplace.

Tax laws that make certain financing alternatives attractive to publicly-owned utilities do not always offer a similar advantage for investor-owned utilities (e.g., tax-exempt bonds coupled with generally high credit ratings result in average lower borrowing rates for the financing of capital improvements for publicly-owned utilities when compared to the use of corporate or taxable bonds used by investor-owned utilities). However, tax laws have provided benefits such as the use of accelerated depreciation that allows for the payment of income taxes included in rates to be deferred which will improve cash flow and the availability of financing capital improvements.

BUDGETING, PLANNING, AND FORECASTING

Comprehensive forecasts of revenues, expenses, and the timing and funding of capital investments assist management with decisions regarding the need for rate increases and effects on customer bills. An effective planning process is fundamental to positive public relations with customers and regulatory agencies.

As discussed in chapter 1, good financing plans begin with a multiyear capital expenditure budget detailing, in part:

- All major construction projects
- Routine or ongoing renewals, replacements, betterments and upgrades of mains, services, pumping facilities, meters, and hydrants
- Purchases of equipment and other general plant materials, such as motors and water-treatment equipment
- Any planned land or water source purchases

Good financing plans also include a corresponding funding analysis detailing the source of financing of the respective multiyear capital expenditure budget and an operating budget for the same timeframe which would include incremental increases or decreases in operation and maintenance expenses due to the implementation of the multiyear capital budget.

These combined comprehensive long-range budgets provide the basis for determining:

- Cash requirements
- Targeted capital structure ratios based on rate-base asset investment
- Appropriate and timely rate increases
- Capital-financing options

GENERATION OF FUNDS

Capital expenditures funds are derived both internally and externally. Internal funds or cash flows are derived from utility system operations and are generally defined as net income, plus deferred taxes, plus depreciation and amortization expenses (considered as noncash expenses), less dividends. External funds, as the name suggests, are funds raised from external sources such as the debt and equity markets and contributions-in-aid-of-construction or customer advances.

Internal Sources of Capital

The rate-making process for investor-owned utilities does allow for the recovery of funds to finance ongoing capital expenditures (referred to as internal sources of capital). Although the actual sources of funds for a specific project depend on a variety of financial circumstances and timing, funds for routine replacements and renewals are normally generated internally from annual utility plant depreciation and amortization expense accruals and the rate of return earned on utility plant investment made by an investor-owned utility, both of which are included as a component of the rates charged for water service.

Annual accruals from depreciation and amortization are a source of capital funding since these are noncash expenses that are recovered in rates and are available for any purpose of the investor-owned utility, including capital investment. Because an investor-owned utility is allowed to earn a rate of return on its allowed rate-base asset investment, such earnings are also available for any utility purpose, including the funding of capital and the payment of dividends to shareholders.

The internal sources of funds available within an investor-owned utility are:

- Annual depreciation accruals
- Accumulated cash (retained earnings)
- Deferred taxes
- Cost of carry-on construction
- Rate of return on investment
- Infrastructure surcharges

Annual depreciation accruals (Asset replacement allowance). Depreciation is an accounting process of allocating the cost expiration associated with the use of tangible property, plant, and equipment against periodic revenue. It represents an expense recognized in the determination of the revenue requirements of an investor-owned utility and thus reduces the pretax income of the utility. Depreciation reflects the amount by which the value of a utility plant is not restored by routine maintenance activities. This loss in value is primarily caused by wear and tear from operation, action of time and other elements, deterioration and decay, inadequacy, obsolescence, changes in demand for service, and changes in regulatory requirements. For one or more of these reasons, each tangible element of a utility must ultimately be retired and, if necessary, replaced.

Investor-owned utilities are permitted to include in revenue requirements depreciation on tangible property that is considered as being used and useful (The appropriate amount of plant to serve current needs plus a reasonable amount of capacity for future needs, without excessive capacity beyond that which can be used during some reasonable future period). The amount of depreciation to be recognized as a revenue requirement may be reduced for several reasons which could affect funds availability. Many jurisdictions reduce depreciation for assets that do not serve existing customers (non-used and useful plant, plant held for future use) and by the amount of utility plant that has been contributed by developers or customers (contributions-in-aid-of-construction or CIAC) such that only the depreciation on the net or equity investment made by the investor-owned utility is recognized. The CIAC can be in the form of a property contribution (e.g., developer-constructed water distribution lines that are contributed to the investor-owned utility) or in the form of a cash contribution (e.g., a capacity or main extension charge, meter installation charge).

The operational life of utility plant and the resulting depreciation rates are approved by utility commissions at levels that permit systematic amortization of the original cost of investment of the various functional components of utility plant (e.g., water production utility plant, water distribution lines, meters, etc.), over the estimated average service or useful lives. Depreciation is a noncash expense; cash generated by recognition of depreciation charges in the water rates is available for use to retire debt or to finance capital expenditures. In many instances, the funding of capital improvements from the annual depreciation accruals are for expenditures that do not usually constitute substantial investments and do not warrant external funding. They typically include replacement of distribution and service lines, meters, pumps, hydrants, and purchase of minor equipment, office furniture, and tools.

While such capital replacement expenditures preserve the overall original total system investment in a utility plant, annual depreciation based on the original cost of the utility plant placed in service may not provide adequate capital to replace the investment during times of inflation. Except for a fully depreciated utility plant that continues to be in-service, a CIAC-related plant, and land, annual depreciation is recognized from all assets in service and represents a return of capital investment to the investor-owned utility. However, all original utility plant investment is not replaced in any single year. As investment is replaced, a new original cost value is included in the determination of annual depreciation accrual and as a component of the rate-base asset investment, on which the investor-owned utility is allowed to earn a return.

If funds received by an investor-owned utility from the receipt of the annual depreciation allowances are not retained and reinvested within the respective utility system, the original utility plant investment may be gradually dissipated. This reduction in utility plant investment will also affect the rate of return earned on rate-base assets since the annual depreciation accrual reduces the amount of such assets and correspondingly the rate of returned earned by the investor-owned utility. In this case, external financing requirements will be greater than otherwise anticipated. It is advisable that investor-owned utilities retain the internal sources of capital where practical to allow for maintenance of their investment in utility fixed assets.

Deferred taxes. Taxes are generally included as a revenue requirement based on the liability that is incurred from a corporate basis. Taxes may include, but not be limited to, property taxes and state and federal income taxes. In many instances, timing differences may result whereby the payment of the current period tax liability is postponed or deferred to future periods; such taxes are referred to as *deferred taxes*. The current period taxes are normally charged against current operating results and recovered in water rates. Since the deferred taxes represent a liability (payable) to the taxing authority or government, the deferred taxes provide another immediate source of capital. Accordingly, cash that may be generated by deferring taxes is also a source of funding for capital improvements.

Construction-work-in-progress interest costs. nother source of internal funding in certain jurisdictions is interest costs incurred to finance construction-work-in-progress (CWIP) before a utility plant is placed in service. A major concern of investor-owned utilities today is the methodology that regulatory agencies use to recognize the interest costs associated with CWIP. Some regulatory agencies exclude CWIP from the rate base on which utilities are permitted to earn a return until that property is placed in service. In such instances, an allowance for funds used during construction (AFUDC) is considered a part of the cost of the property and is capitalized, along with the plant investment, and may be eligible to become a part of the rate-base assets once the utility plant is placed into service. AFUDC includes the net cost for the period of construction of borrowed funds used during such period at a reasonable rate applied to the respective funds used, as allowed by the regulatory agencies. In such instances, it is important for the investor-owned utility to delay debt repayments until the CWIP is completed and accounted for in the rate-base assets since both the applicable depreciation expense and corresponding rate of return on the rate-base assets can be earned and applied to the debt repayment.

Because of the unpredictable and erratic nature of inflation rates and interest costs, investor-owned utilities have sought relief from the administrative burden of charging interest during construction by requesting the inclusion of CWIP in their rate base. This permits earnings at the overall allowed rate of return during the construction period instead of when the utility plant is placed into service. Many regulatory agencies now include at least some CWIP (for a certain period beyond the end of the test year or that which is nonrevenue producing) in the rate base and permit the investor-owned utility to earn a return on this CWIP.

On very large projects, an investor-owned utility may work with outside investors specializing in project finance who will provide a turn-key constructed facility. The investor-owned utility specifies its needs and buys the completed facility from the outside investors when it is placed into service. The investor-owned utility can immediately place the asset on its books and earn a return. This approach may cost the utility a little more for the asset than if it constructed the asset itself. However, the utility's funds are not tied up in CWIP without the ability to include such costs in its rate base.

Return on investment. Total return on investment for an investor-owned utility is designed to recover the cost of using various sources of internal (equity) and external (debt) funding. The return that an investor-owned utility is allowed to earn is predicated on the amount of rate-base assets that is recognized or allowed by the regulating agency and the rate of return that is often equal to the utility's weighted average cost of capital to be earned on such utility plant investment.

The rate of return is intended to allow an investor-owned utility to meet its obligations to capital investors (stockholders and lending institutions) and to compete on reasonable terms in the financial markets for future capital funds. Total cost of capital primarily includes the following components:

- Capital secured by the issuance of long-term debt
- Capital secured by the issuance of preferred stock
- Capital secured by the issuance of common stock and common stock equity

The cost of long-term debt and preferred stock are usually established contractually and can be determined from investor-owned utility or corporate records. The cost of equity capital is frequently a controversial subject for the utility, consumer, and regulatory agency. Ideally, the cost of equity capital is equal to the return that investors will require on their market-valued commitments to continue committing their capital resources to an enterprise. Because investors have different views and expectations about the return to be realized on their investments, determining the market rate of return to attract capital is difficult and typically subject to special studies, including expert testimony, that recognize market conditions.

The return earned on its permitted rate-base investment coupled with the recovery of depreciation through rates provides a source of funds to finance ongoing capital infrastructure and pay the cost of long-term debt issued to finance the infrastructure needs and provide dividends to investors. Refer to state administrative codes and rate regulatory guidelines for information about the expenditures and the basis for determining the amount of rate-base assets that may be included in investor-owned utility water rates.

Infrastructure surcharges. Some state regulatory commissions have approved the use of infrastructure surcharges by investor-owned utilities, which permit recovery of a rate of return on, and the recognition of depreciation on, certain qualifying utility plant investment between rate cases. Although the details of each surcharge vary from state to state, all permit the recovery of a return and depreciation, which is applied as a percentage to metered customers' bills. Certification that an investor-owned utility is not earning returns in excess of its last authorized rate of return is a prerequisite to implementation of the surcharge. The surcharge is usually capped (generally ranging from 5% and 9%

between rate cases, with some states having annual caps as well) and surcharge revenue is included as a component of the existing rate revenue in the investor-owned utility's next rate case, at which time the surcharge is reset to zero. In many instances the funds received from a surcharge are restricted to a particular use consistent with its purpose and are not considered as a general-purpose revenue. A typical infrastructure surcharge formula is

$$\text{Infrastructure surchange percentage} = \frac{(\text{Rate of return} \times \text{Qualifying plant}) + \text{Depreciation}}{\text{Total rate revenue}}$$

In this formula, *qualifying plant* refers to the capital facilities or assets that are allowed to be included in the surcharge between rate cases, such as buried infrastructure replacements.

External Sources of Capital

Typically, major capital projects cannot be financed from funds generated from normal annual operations by an investor-owned utility because of the magnitude of expenditures associated with these projects. For major construction projects, external sources of capital (debt and equity issues) generally must be secured. These sources include:

- Debt
  - Long-term debt
  - Short-term debt
- Equity
  - Common stock
  - Preferred stock
- Contributions in aid of construction

Debt. Bonded indebtedness is normally secured by a mortgage or conditional lien on the investor-owned utility's property. It has the highest payment lien, or call, on the revenues derived from operations after the payment of operations and maintenance expenses and is ultimately secured by the utility plant investment of the investor-owned utility. That is, interest payments on bonded indebtedness must customarily be paid before dividends can be paid on preferred and common stock. In case of liquidation, bonded indebtedness principal is repaid before repurchase of outstanding stock.

From a bond investor's viewpoint, bonded indebtedness is the most secure and least risky form of capital investment in an investor-owned utility. Interest on debt is also deducted for income tax purposes. Accordingly, the cost of this debt is generally lower than that the cost of preferred or common stock. If the company is financially healthy, unsecured notes may also be issued to fund capital expenditures. Unsecured debt has a secondary or subordinate priority to debt secured by a mortgage or lien. Many state regulatory commissions require that any long-term (greater than 12 months) securities issuance receive approval prior to issuance.

Long-term debt. Long-term debt is generally issued by investor-owned utilities for terms ranging from 10 to 20 years with the debt being classified as either taxable (interest received by investor being taxable) or tax-exempt in cases involving private activity bonds or industrial development bonds (interest received by investor being nontaxable). The terms of the debt obligation will influence the borrowing rate for the debt and will ultimately affect the weighted cost of capital calculation (rate of return earned which is used to pay the interest expense on the debt). In most cases, taxable debt is privately placed with an insurance company or other large institutional investor. From time to time, utilities may prefer to issue debt by means of a public offering to maintain relations with retail investors.

Investor-owned utilities in many states also have access to tax-exempt financing (e.g., industrial development bonds) if the state in which they operate is willing to designate some of its capital allocation to the utility. Investor-owned utilities may also be eligible to apply for state revolving fund (SRF) loans, another source of low interest financing that combines state and federal funds for projects that improve water quality. Funding for the SRF loans is dependent on a capital project priority ranking and the number of other publicly-owned and investor-owned applicants competing for the available funds each year.

Historically, the average annual interest rate or cost of a long-term (i.e., 30-year) tax-exempt issue is approximately 2.0 to 3.0 percent lower than the cost of a comparable taxable issue. To the extent that the tax-exempt market is available to an investor-owned utility, many utility commissions and rate-case interveners (large users and public advocates) encourage this method of financing because of the substantial cost savings over taxable debt. An investor-owned utility is subject to rules promulgated by the local regulatory body, as well as to previous indenture language, corporate resolutions, and language in the utility's by-laws.

Long-term bonds are generally secured by a first mortgage or lien on the utilities' or issuer's property. Some utilities that do not have a large amount of secured debt have successfully issued debentures, which are a legal contract between the utility and the creditor bondholders, that spells out all the detailed terms and conditions of the loan; these types of loans are usually subordinate to the first mortgage bonds. Historically, most long-term fixed-rate issues had been marketed with bond insurance that can further decrease the cost of financing. However, in the current market environment the ability to obtain bond insurance is very limited, resulting in the need for other revenue and bond surety enhancements, including the funding of a debt service reserve fund (typically equal to the maximum annual debt service payment), use of alternative revenue pledges for the repayment of the debt, restriction of company liquidity as a surety toward debt repayment, issuance of "closed-end" loans to increase creditor protection, and other options. Issuance costs for tax-exempt debt are much higher than for a private placement because of underwriters' fees and additional counsel costs to represent underwriters and to opine on the tax-exempt status of the debt.

Short-term debt. Short-term debt is often used to finance capital projects during construction to limit interest costs and to more accurately match cash flow requirements. This debt may be in the form of variable-rate bank lines of credit, fixed and variable-rate short-term bond anticipation notes, commercial paper programs, and other debt instruments that are considered to have a limited duration (generally not exceeding 5 years). Water utilities usually replace these short-term debt instruments with long-term, fixed-rate debt, or equity, or both, as soon as the completed projects are ready to be placed in service, which will allow the investor-owned utility to then earn a rate of return on and to begin to depreciate its investment in the constructed utility plant.

Equity. Equity capital may be provided by issuing common or preferred stock which represents ownership in the utility by the investor (not a liability for payment of a borrowing). Preferred stock has a preference or priority over common stock, which may relate to dividends, assets, redemption, and convertibility.

Common stock. Common stock has value since it confers a claim to the utility's net income, including growth in the utility, and a claim to any asset value that may remain after all creditors and preferred stockholders have been paid. An investor-owned utility issuing new common stock must ensure that that two primary objectives are met: (1) the dilution of earnings per share experienced by existing common shareholders due to this issuance is minimized as much as possible; and (2) adequate funds will be available to pay common dividends to both existing and new shareholders after the issuance of such stock.

To offset the dilutive effect of issuing new shares, most common stock issues occur in conjunction with an acquisition or a rate case that increases earnings. Many investor-owned water utilities also have dividend reinvestment plans or customer stock purchase plans, or both, that can satisfy moderate capital requirements throughout the year.

Preferred stock. Preferred stock is the second least-risky type of investment. Investors in bonds and preferred stocks sacrifice most of the benefits that may ensue from a utility's future growth. In exchange, bondholders get a promise by the investor-owned utility to pay a predetermined amount of interest on specified dates and repay principal at a specified time. Preferred stock foregoes the benefit of growth but does get a prior claim on income and assets ahead of the holders of common stock. Accordingly, preferred stock dividends and, in case of asset liquidation, its principal are payable before such payments are made to holders of common stock. This hybrid of debt and equity is generally considered comparable on a risk basis to a junior lien bond. Thus, its cost is higher to the investor-owned utility than debt, but not as high as common stock.

Contributions-in-aid-of-construction. In certain cases, external funding for construction is obtained by direct contributions from those customers who benefit from the construction or from the availability of capacity. For example, developers usually construct and donate, or pay for the on-site water distribution mains as well as off-site transmission main extensions to their developments. Investor-owned utilities may charge plant capacity fees or system development charges, main extension fees, service line and meter installation fees, and other forms of capital charges to recover the cost of utility constructed or installed facilities. The receipt of property (utility plant) and/or the reimbursement payments must be accounted for in the form of contributions-in-aid-of-construction (CIAC) or as customer-advances-for-construction (CAC). Although such contributed capital does provide a source of funds or benefit to the investor-owned utility, such amounts are not considered as equity capital and usually will be deducted from the determination of rate-base assets that form the investment basis to earn a rate of return by the utility.

CIAC represents any cash, services, or property permanently contributed by customers to the investor-owned utility for construction purposes or for the allocation of capacity in such facilities. CACs also represent contributions from developers or customers that may provide benefit to other properties in addition to the contributor (e.g., a line extension that can serve many parcels). In many instances, and as a condition of constructing and financing the asset, the utility and developer will enter into a reimbursement agreement that will require the utility to refund part or all of the construction cost to the developer over a specified time period as other customers are connected to the system and pay for their pro rata share of the contributed asset. Any unrefunded balance remaining at the end of the term of the reimbursement agreement period is transferred to the CIAC account.

Although it is possible for an investor-owned utility to benefit in the short term from customer growth (via operating revenues) while not directly financing any significant additional capital investment, this is normally not the case for the long term. Generally, regulators do not allow investor-owned utilities to benefit from the receipt of contributed capital to the utilities through depreciation or from earning a return on undepreciated contributed assets. Investor-owned utilities generally can only receive a return of their invested capital (investment specifically made directly by the investor-owned utility) and earn a return on their outstanding undepreciated investment. Over time, the investor-owned utility will be confronted with replacing the contributed plant and may need to increase water rates to reflect the need for the utility-financed additional capital investment requirements during the long term.

Utility Plant Replacement Policy

Management should pursue a replacement policy that minimizes total long-term costs and is consistent with rendering safe, adequate, and reliable water service. However, in developing an asset replacement plan, detailed asset replacement cost and performance records regarding asset condition and remaining useful service life may not always be available for evaluation. Consequently, awareness of the total cost-control objective on the part of operating personnel and the financial manager is beneficial for planning purposes. The annual cost for normal replacements should be incorporated into capital budgets and cash-flow projections. Because of the long service life of water system plants and equipment, most facilities are replaced infrequently. However, delayed replacement can become an additional problem when costs for maintaining an aged plant become excessive.

Provisions for replacements of major facilities must be incorporated into the financial and rate-setting planning process. When preparing the utility capital plan, investor-owned utilities incorporate the regulatory lag in the establishment of rates, which would also include capital investments as a component of the rate-base assets. *Regulatory lag* generally refers to the timeframe that transpires from when investor-owned utility plant assets are financed, constructed, and placed into service and when those assets are included in water rates. The constructed utility plant is usually included as a revenue requirement to be recovered from rates only after completion of a formal rate adoption process or proceeding before state regulators, which could require as long as 24 months to complete. The investor-owned utility needs to balance cash flow and overall funds availability while recognizing cost recovery issues; major capital investments that constitute a significant commitment of financial resources generally require funding from external sources.

ATTRACTING CAPITAL

A primary objective of investor-owned water utilities is to attract capital at the most cost-effective rates to minimize the impact on the customers' water service rates. To attract capital, regulated utilities must be vigilant of their revenue needs and file for rate relief as frequently as necessary through evaluating long-range budgets and capital plans. An investor-owned utility that presents itself to be progressive in maintaining adequate rates and earning its allowed rate of return on rate-base assets will be more successful in the capital marketplace. To attract capital at minimum cost, the investor-owned utility must also use a balanced capital structure. The most balanced and efficient capital structure combines debt and equity capital.

Capital Structure

Investor-owned utilities' capital structures reflect their mix of debt and equity funding of utility rate base. Each company's capital structure is influenced by its individual circumstances, including the size and intensity of its capital improvement program, the regulatory environment, the magnitude of CIAC and CAC, and the impact of financing alternatives on customers' water service rates. Many state regulatory agencies (public service or utility commissions) establish a target capital structure or basis for determination of the equity component for use in calculating the weighted average cost of capital and a utility's rate of return. An example of typical capitalization ranges for investor-owned water utilities in the United States is shown in Table 7-2.

Rating Agency Considerations

The size and magnitude of a utility's capital plan often determine the prudence of obtaining a corporate credit rating or a debt rating from one or several rating agencies. Seeking a rating from all the rating agencies, or even just one, is time consuming and can be costly. Frequently, the investor-owned utility's investment banking firm will provide advice on the economics and the prudence of obtaining one or more ratings.

Regulatory Approvals

Many investor-owned utilities must file with and obtain approval from its regulatory agency and may be required to obtain approval from the Securities and Exchange Commission before issuing long-term debt. Primary financing documents, such as the bond or note purchase agreement and the supplemental indenture, board of directors' resolutions, and supporting financial exhibits justifying the purpose and selection of the particular financing, may be required to be filed with the regulatory agency for approval before the actual financing. The financing's prudence, purpose, and impact on the investor-owned utility's capital structure must be demonstrated and justified to the regulatory agency to gain the needed support. Many regulatory agencies have their own guidelines for what is considered an appropriate capital structure for investor-owned utilities under their jurisdiction.

Regulatory Lag

A major hurdle in keeping a water utility attractive to potential investors is regulatory lag—the lag between the time when a rate case is filed and the time when the regulatory agency provides rate relief or approval. While state regulatory agencies have done much in recent years to accelerate consideration of rate relief requests and to alleviate the impact of rate approval lag, a degree of lag cannot be avoided. Therefore, investor-owned utilities must maintain vigilance over their revenue needs and file for rate relief as frequently as necessary. A strong basis for rate adjustment and limiting the complexity of the rate filing will generally reduce the time for the rate proceeding. An investor-owned utility that is progressive in maintaining adequate rates will be more successful in the capital marketplace.

To determine revenue needs, most regulatory agencies require a rate filing to be based on statistics that show historical test-year (the year which serves as the basis for the rate determination) expenditures, but often allow for modifications for known and measurable changes. Some regulatory agencies provide for fully projected test years in the determination of rates. Capital markets, however, consider the future likelihood of repayment in their credit evaluations of financial strength and rate sustainability. Projected test years can allow an investor-owned utility to be better prepared to develop a financial plan that will be meaningful to both regulatory agencies and capital providers.

To reduce the impacts of regulatory lag, many regulatory agencies now allow investor-owned utilities to file for a price index rate adjustment based on a specific formula for the price index. The index process would be in lieu of a formal rate filing and generally

Table 7-2 Typical investor-owned utilities capitalization example

Type of Capital	Maximum, %	Average, %	Minimum, %
Long-term debt	60	50	33
Preferred stock	4	2	0
Common stock (plus retained earnings)	65	48	35

allows for the recovery of inflationary impacts on the cost of operation and maintenance. This allows the investor-owned utility to marginally increase rates to maintain operating margins without the need for a costly and time-consuming formal rate application. Additionally, many jurisdictions allow for a limited proceeding rate filing that addresses a specific situation or condition and is not considered as a formal rate filing. In many instances, the reason for a limited proceeding would be to adjust the level of rate-base assets and operating expenses (depreciation expense) for a significant capital addition made by the utility; no other aspect of the utility operations, including the capital structure or rate of return, would be evaluated by the regulatory agency in a limited proceeding. Generally, the regulatory lag for a limited proceeding is shorter because of the reduced complexity in the rate evaluation and approval process; many jurisdictions will place a cap on the total rate adjustment that may be approved by use of a limited proceeding.

OTHER CONSIDERATIONS

Other factors determining the attractiveness of an investor-owned water utility's stock, which impacts the investor-owned utility's ability to secure low-cost external financing, are the dividend yield, the company's record of earnings and dividend increases, the dividend payout ratio (which indicates the future growth potential of the dividend), and potential earnings growth of the utility. These factors generally relate to the overall financial sufficiency of the utility and are a measure of financial strength and management. As long as the investor-owned utility maintains its rate of return, funds an ongoing capital reinvestment program financed with a proper balance of equity and external capital, and provides investors a long-term return on its investment, it should be able to secure capital at favorable levels and maintain long-term sustainable rates for service.

Balancing Investor and Customer Requirements

The interests of stock and bond investors must be balanced with the interests of an investor-owned utility's customers. Investors need to feel secure that the utility has a sound financial position and that a reasonable return will be earned on their investment commensurate with the risk associated with that investment. Customers, however, are interested in receiving safe, adequate, and reliable service at reasonable rates. Customers must be informed that the investor-owned utility's ability to continue to provide such service depends on its maintenance of both a sound financial position and a reliable physical plant, which are also necessary to attract needed capital at reasonable rates.

SUMMARY

Options available for financing differ between publicly-owned and investor-owned utilities. This chapter has discussed considerations specific to investor-owned utilities, including the planning process, internal and external sources of capital, attracting capital, and balancing the interests of both customers and investors.

BIBLIOGRAPHY

AWWA. 2012. Manual of Water Supply Practices M1, *Principles of Water Rates, Fees, and Charges*, Sixth Edition. Denver, Colo.: American Water Works Association.

Bonbright, James C. Albert L. Danielsen, and David R. Kamerschen. 1988. *Principles of Public Utilities Rates*, Second Edition. Arlington, Va.: Public Utilities Reports, Inc.

Phillips, Charles F. 1993. *The Regulation of Public Utilities, Theory and Practice*, Third Edition. Arlington, Va.: Public Utilities Reports, Inc.

This page intentionally blank.

Chapter 8

Special Considerations for Canadian Utilities

Utilities around the globe strive to find the optimal means to finance capital projects. This chapter discusses some of the aspects of utility capital financing in Canada, including planning for operations (the operating budget), long-term assets (the capital budget), and the advantages of setting policy to ensure fiscal stability. Policy matters in this chapter focus on medium-term stability targets, cross-generational and reserve funding, rate stability and cost of service, and capital financing. Also discussed are the benefits of public communication and citizen engagement and benchmarking.

FINANCIAL AND CAPITAL PLANNING PROCESS

Revenue Requirements

For utilities, revenue requirements can include costs associated with operations and maintenance (O&M), taxes, payments in lieu of taxes, debt service, nondebt financed capital improvements (often referred to as pay-as-you-go capital), and funding of specific reserves. In Canada, revenue requirements may also consider a dividend payment to the municipality as a return on investment in the utility. Similar dividend-type payments can be found within the United States, more commonly in the form of a franchise fee.

The two basic approaches for establishing revenue requirements are the utility or *equity-based* approach and the *cash-needs* approach. The equity-based approach is more directly tied to accounting principles and provides for a rate of return on investment, while the cash-needs approach generally focuses on recovering the annual cash requirements of the utility. Typically, publicly owned Canadian utilities follow the cash-needs approach. One advantage of this simpler approach is that management is able to develop a rate strategy that directly recovers planned cash requirements. The cash-needs approach can also be more intuitive and the methodology is easier to communicate with internal and external stakeholders. Regulated utilities in Canada typically follow a cost-of-service framework established by their regulatory bodies for determining revenue requirements, which is similar to investor-owned utilities in the United States.

Operations Planning

Operating budgets identify the budgeted operating costs of the utility and are used to identify the revenue that is needed to recover on-going service costs and plan for the maintenance of future capital projects. Operating budgets for Canadian utilities are typically approved on an annual basis by regulating authorities, municipal board, or a board of directors and can include, for planning and information purposes, a number of forecasted years beyond the current budget year. Recently, it is becoming more common to approve a three-year operating budget along with a multiyear financial plan as a number of Canadian cities have adopted a longer-term, more stable approach to future spending. Approval of operating budgets beyond the traditional one-year period drives continuous improvement and requires thorough planning to maintain financial stability and still provide the operational support needed to meet existing, planned, and emerging capital requirements.

Capital Planning

Capital planning usually spans multiple years. Some Canadian provinces require submission of capital improvement plans (CIPs) spanning multiple fiscal years as a condition of infrastructure financing. For example, in Nova Scotia, a CIP spanning 5 or 10 years must be submitted online annually to the Canada Nova Scotia Infrastructure Secretariat. While many factors drive capital investment, such as increasing the level of service, growth, and economic development, successful capital planning not only utilizes annual capital budget processes, but also considers the combination of regulatory requirements and asset management.

Regulatory requirements are one driver of capital planning. For Canadian public utilities, provincial and federal bodies representing health and water stewardship regulate the quality of drinking water using established criteria (e.g., Government of Manitoba 2007 and Government of Canada—Health Canada 2012). A provincial regulatory body can perform an annual compliance audit that provides details of the licensing requirements and the water system performance as it relates to the terms and conditions set out in the utility's operating license. Affected capital infrastructure and associated operating costs influence revenue requirements, necessitating longer-term planning to ensure utility rates remain at affordable levels while still providing the required funding needs.

Many Canadian utilities have adopted asset management as a way of doing business that ensures wise investment in resources over the long term by continually balancing the delivery of an established level of service at an acceptable risk, while minimizing the life-cycle cost of ownership. Some Canadian utilities have further introduced an investment planning (IP) process as an additional capital planning tool. The IP framework can be summarized as follows:

- **Goals**: Define level of service
- **Needs**: Identify improvements to level of service
- **Solution**: Present business case for team evaluation
- **Priority**: Prioritize and select project(s)
- **Plan**: Develop asset management and financing plans

Clearly identified levels of service help balance corporate and customer requirements with technical requirements. For example, the provision of safe drinking water is balanced with the number of interruptions to customer water service, while minimizing technical outages resulting from water main breaks. A risk approach is often used to identify capital maintenance priorities, employing a business case (BC) evaluation process that includes detailed risk and condition assessments. The BC process assists in ensuring a best-value capital portfolio is selected for funding. The process includes evaluating the financial, social, environmental, and political impacts and the criticality and risk components and ensures that engineering, operations, and financial perspectives are all considered. The aim of the BC process is to consider a range of approaches that best meets the needs and provides value for ratepayers through the selection of the lowest whole-life cost solution. Business cases play an important role in the investment planning process by defining requirements, evaluating options and solutions, and documenting costs, issues, and constraints. The successful BCs are prioritized and used in developing the multiyear CIP.

FISCAL POLICY

Many Canadian utilities, along with their counterparts in the United States, are self-supporting, and establishing fiscal policy is sound business practice. Utilities provide an essential service and an uninterrupted supply of safe drinking water. This is not only a goal but a necessity. Given such a requirement, Canadian utilities must balance this often costly, essential service with a sense of business strength. Financial feasibility for the utility is maintained by establishing policies, such as determining and setting realistic and attainable fiscal stability targets, preparing for funding major capital projects over the useful life of the asset or on a pay-as-you-go basis with reserves, identifying opportunities associated with changes in utility rate structures, and acquiring unique capital project financing.

Medium-Term Stability Targets

Canadian utilities employ fiscal targets or policies to provide medium-term stability in the event of unforeseen circumstances. Stability targets focus on ensuring liquidity in the event there is an interruption in rate funding, increased capital costs due to new regulatory requirements or a short-term funding emergency where there is not enough time to secure a borrowing agreement. Setting and maintaining stability targets assists the utility in meeting both the current needs of customers and unanticipated events. One way to measure short-term stability is through a working capital reserve target. Working capital reserve targets can vary among utilities and can be based on a number of factors. For example, a utility subject to extreme weather events, either drought or torrential rains, might need a higher working capital target than a similar utility located in a more neutral weather zone. Medium-term stability targets for utilities are common business practice in Canada and the United States where utilities reference best practices, including USEPA's Effective Utility Management program that was established in 2008 as a partnership among AWWA, the Water Environment Federation, the National Association of Water of Water Companies, and other organizations (USEPA 2008). For example, a minimum working capital balance of 60 to 90 days of O&M expenses is a commonly used target.

Reserves Policy

Because of the capital-intensive nature of water utilities, the use of reserves helps prepare for major projects in the years prior to the actual need for the funds, balancing current rate funding with longer-term debt funding. For example, the construction of a large water treatment facility is often a high-cost, multiyear endeavor, involving a number of funding sources. The establishment of a dedicated water treatment plant reserve can be a wise business decision, where the accumulation of funds prior to the actual start of construction combined with long-term funding during asset completion, offers the avenue for cross-generational responsibility. The policy can direct a portion of revenue to accumulate in the reserve over a set number of years and for a set amount representing the current rate-payer's contribution to the benefit of the asset. Long-term debt could account for the remaining funding requirements and would be the responsibility of future rate payers.

Reserve policies in Canada can be specific in nature or can be for a general purpose. While practiced in the United States, rate stabilization reserves only recently are being introduced in some Canadian utilities. Rate stabilization reserves are dedicated to help mitigate unexpected pressure on utility rates and avoid rate shock. In the event of unexpected, substantial capital or operating costs, the shock of sharply increasing rates to meet the funding requirements is offset or mitigated by drawing on the resources of a rate stabilization reserve.

Rate Stability and Cost-of-Service

Increasing capital costs largely driven by Canadian regulatory requirements, in addition to decreasing water consumption patterns, have driven the need for policy that ensures equitable and stable rates. One way Canadian utilities have planned to meet rate stability is to introduce the cost-of-service (COS) concept. The introduction of COS rates help to institute a policy that supports established methods of funding services and provides a defensible correlation between what customers pay for and the benefit they receive (AWWA 2012). A COS rate methodology, while more common in parts of the United States where the availability of potable water is not as abundant, is becoming more prevalent in Canada. The introduction of a COS policy and resulting rate structure assists in funding large, costly capital projects by assessing who will benefit from the new infrastructure and by how much, and ultimately allocate those costs to the appropriate user group. In addition, with such comprehensive rate development analysis, multiyear rate approvals are more readily accepted by governing bodies. Multiyear rate approval supports transparency and consistency for customers and financial stability for the utility.

Financing Capital Projects

The need for utilities to be able to provide reasonable rates, adhere to regulatory requirements, maintain system infrastructure, sustain financial stability, and meet the needs of customers is driving the pursuit of innovative approaches for capital financing in Canada. Common financing alternatives can include short- and long-term debt, government programs and grants, and private sector investment.

Debt financing of capital projects within Canadian utilities is similar to those of the United States. Generally, Canadian publicly owned utilities are governed by a municipal or provincial charter where the municipality is authorized to incur long-term debt and issue debentures. Long-term debt is typically issued for capital purposes only. A financial management plan will provide the avenue for guiding financial decision making, which supports a fiscal strategy with a sustainable long-term approach. Debt management policies detail the parameters for issuing and managing outstanding debt and set forth debt limitation ratios and conditions to be followed by the utility and remaining tax-supported departments. For example, for some Canadian publicly owned utilities, debt is issued under the municipality's name and credit rating, and investors rely on the consolidated financial position. Borrowing in British Columbia is pooled under the Municipal Finance Authority and in Alberta by the Alberta Finance Capital Finance Authority.

Alternately, for some privately owned utilities, debt is issued as an obligation of the utility and does not carry the overall municipal guarantee. Debt issues within Canada vary depending on ownership and can include but are not limited to: bullet bonds or sinking fund issues with a fixed coupon rate; serial structure typically used for shorter term and smaller amounts; amortizing loans, while not common, provide the ability to customize principal and interest components; or bank loans, common for smaller capital projects of odd terms.

As throughout North America, favorable interest rates for Canadian utilities depend on many factors, including the level of comprehensive financial planning, dedicated revenue streams, and resulting agency credit ratings. Within Canada, the opportunity also exists for utilities to borrow at a rate lower than the market rate. For example, in recognition of the need to support infrastructure, a Municipal Infrastructure Lending Program (MILP) was introduced by the Government of Canada that provides low-cost loans to municipalities through the Canada Mortgage and Housing Corporation (Government of Canada MILP 2009). These low-cost loans decrease the cost of borrowing for municipalities and can be used to directly invest in municipal housing-related infrastructure. Eligible municipal infrastructure projects must directly relate to housing, contributing to the efficient functioning of residential areas. Projects would include, for example, infrastructure related to the provision of water or wastewater. Loan terms are typically 15 years with consideration given to longer-term requests and subject to a loan agreement or debenture. Interest is set at the time of the loan by the Canada Mortgage and Housing Corporation.

The Canadian Strategic Infrastructure Fund (CSIF) provides another avenue for funding of water capital projects (Canada CSIF 2011). CSIF is a tri-partied agreement with the federal, provincial, and municipal governments. Funding is available for qualifying, large-scale infrastructure projects in support of sustaining the economic growth and enhancing the quality of life for Canadians. The Canadian government can contribute up to a maximum of 50 percent of total eligible costs.

Strategic relationships between government and private entities in Canada can provide an innovative approach to meeting resource challenges and customer expectations. One difference between municipal financing in Canada and the United States is that municipal debt in the United States is typically tax-exempt, and in Canada it is taxable for all Canadian issues purchased by investors. This makes the cost of private financing more competitive (from an interest rate perspective) with public financing, and thus a more attractive environment for public–private partnerships (PPP).

PPP can be described as a business venture funded or operated through the partnership of government and private sector. As per the Canadian Council for Public–Private Partnerships (CCPPP), in the Canadian content, a PPP relates to the provision of public services or public infrastructure and necessitates the transfer of risk between partners (CCPPP 2011). The CCPPP has developed *Public–Private Partnerships: A Guide for Municipalities* to assist in delivering Canadian infrastructure and public service solutions. The PPP agreement offers the sharing of risk and will dictate the parameters in which the parties, municipal government, and private sector provide the service and to what degree each party assumes financial, technical, and operational risk. For example, under a private finance initiative agreement, capital investment may be made by the private sector on the basis of a contract with the municipal government to provide agreed services, and the cost of providing the service is borne wholly or in part by the municipal government. The PPP framework helps all parties to control the dynamics leading to cost overruns and delivery delays, which often occur under traditional capital project management.

OTHER CONSIDERATIONS

Communication and Public Consultation

Strategic management, financial management, and project management, along with traditional capital financing and innovative partnership funding approaches, are part of the capital financing decision-making process in Canada. One often underestimated or underutilized consideration is the opportunity that public consultation and outreach can provide during the financial planning and capital financing process. A coordinated and consistent approach to public or citizen engagement is a developing practice in Canada (National Collaborating Centre for Methods and Tools 2011). Components of a public outreach can include:

- Public communication: An avenue to share and disseminate information from complex and often hard-to-find government resources to the public that may be affected. Information can include administrative reports, requests for proposals, proposed service changes, and community events, and can be conveyed through a variety of mediums such as radio, web sites, print, or face-to-face.
- Public consultation: Specific policy or service issues could benefit from public consultations, facilitating a better and more acceptable end product. For example, municipal representatives can request public input on the issue of replacing an old, manual water-meter reading service with something new, such as automated meter readings. Public consultation methods can include hearings, focus groups, or surveys.
- Public participation: Dialogue between government and the public promotes the exchange of information and provides the opportunity to take often opposing opinions and convert them into defensible conclusions.

A goal of public consultation is to improve communication channels between stakeholders and the government. In Canada, for example, some municipalities have undertaken a consultation process to engage citizens in shaping the direction of the capital budget. Using online surveys, the process requests the public to answer specific questions regarding the capital budget (for example, what the priorities of the municipality should be and what choices would be made if they were in charge of the capital budget). Citizens can also provide feedback through public workshops, written submissions, or e-mail.

Benchmarking

Benchmarking is an evolving practice within Canadian public sector utilities particularly as it relates to financial planning and capital financing. The benchmarking exercise provides valuable information to help assess fiscal performance, the needs of customers, and provide the tools to help support optimum performance. The National Water and Wastewater Benchmarking Initiative (NWWBI) has become accepted standard for Canadian water utilities to report and compare fiscal goals with other utilities. The annual NWWBI Public Report provides an overview of the performance of participating utilities.

Benchmarking data provides:

- The opportunity to compare fiscal performance and operations metrics with other participating municipalities
- The ability to match utility goals and objectives with performance measure indicators

- An avenue for measured, continuous improvement

Benchmarks or goals can be based on the following performance indicators:

- Capital financing: Borrowing from provincial agencies, issue bonds, utilize reserves
- Distribution: Water pipe age, length, size, material type
- Reliability: Capital renewal (length of main replaced)
- Capital costs: New capital investment
- Treatment: Age of treatment plant, year since last major upgrade

Utility performance improvement is a continuous succession of monitoring trends in key business functions within and among utilities as well as the implementation of action plans. Benchmarking is proving to be a useful tool to assist managers to identify opportunities and take proactive steps to avoid and resolve financial and performance issues.

SUMMARY

Canadian utilities face similar capital financing challenges as their North American neighbors—to ensure operating costs and capital planning management result in a strategy that sustains the utility and provides optimal and equitable service to customers. Typically, publicly owned Canadian utilities follow the cash-needs approach in determining revenue needs. The establishment of fiscal policy provides the platform for utilities to balance costly capital infrastructure and associated maintenance costs with business strength. Fiscal policy can include setting medium-term stability targets, establishing reserves and intergenerational equity funding plans, and rate stability. Financing alternatives can include short- and long-term debt, customer or developer contributions, and government grants. The pursuit of innovative approaches for funding capital projects beyond that of traditional debt supports partnerships between government and the private sector. Such partnerships help to control the dynamics leading to cost overruns and delivery delays, which often occur under traditional capital project management.

A coordinated and consistent approach to public or citizen engagement in the capital financing process is a developing practice in Canada. Public consultation can improve communication channels between stakeholders and governments, allow for sharing of decision making, and ultimately result in a service or policy that meets the needs of all parties. The use of benchmarking can provide the avenue to measure, report, and compare valuable information related to fiscal performance measures—tools needed for continuous improvement.

REFERENCES

AWWA. 2012. Manual of Water Supply Practices M1, *Principles of Water Rates, Fees, and Charges*, Sixth Edition. Denver, Colo.: American Water Works Association.

The Canadian Council for Public–Private Partnerships (CCPPP). 2011. Public–Private Partnerships A Guide for Municipalities. http://www.p3canada.ca/~/media/english/resources%20library/files/p3%20guide%20for%20municipalities.pdf. Accessed July 2, 2014.

Government of Manitoba. 2007. Manitoba Drinking Water Safety Act-Drinking Water Safety Regulation. http://www.gov.mb.ca/conservation/waterstewardship/odw/reg-info/acts-regs/040-d101.07.pdf. Accessed July 2, 2014.

Government of Canada—Health Canada. 2012. Guidelines for Canadian Drinking Water Quality. http://www.hc-sc.gc.ca/ewh-semt/alt_formats/pdf/pubs/water-eau/2012-sum_guide-res_recom/2012-sum_guide-res_recom-eng.pdf. Accessed July 2, 2014.

Government of Canada—Infrastructure Canada CSIF. 2011. Canadian Strategic Infrastructure Fund. www.infrastructure.gc.ca/prog/other-autres-eng.html#csif-fcis. Accessed May 30, 2014.

Government of Canada—Infrastructure Canada MILP. 2009. Municipal Infrastructure Lending Program. Canada Mortgage and Housing. www.infrastructure.gc.ca/media/news-nouvelles/2009/20090720fcm-eng.html#l5. Accessed May 30, 2014.

National Collaborating Centre for Methods and Tools. 2011. Engaging citizens for decision making. Hamilton, Ont.: McMaster University. www.nccmt.ca/registry/view/eng/86.html. Accessed May 30, 2014.

NWWBI. Public Report. Canadian National Water and Wastewater Benchmarking Initiative. http://www.nationalbenchmarking.ca.

US Environmental Protection Agency (USEPA). 2008. Effective Utility Rate Management, A Primer for Water and Wastewater Utilities. http://water.epa.gov/infrastructure/sustain/upload/2009_05_26_waterinfrastructures_tool s_si_watereum_primerforeffectiveutilities.pdf. Accessed May 30, 2014.

Appendix A

Applying Risk Assessment to Asset Management

Objective: Manage the risk of asset failure in the water system by prioritizing asset maintenance and capital replacement projects based on the likelihood and consequence of failure of each asset or group of assets of the utility. Some assets pose high risks and need to be identified and managed much more closely than those that pose low or acceptable risks.

Terms: Asset risk, business risk, exposure, and criticality have all been used at various times; however this appendix uses the terms *risk, likelihood* (of failure), and *consequence* (of failure to the utility and its customers) in risk assessment.

Methodology: Applying risk assessment to the management of assets is a long-term process. Although desirable, risk assessment does not require an accurate picture of the current condition of system assets. If the likelihood of failure is currently unknown, start with the consequences of failure and let these drive the condition assessment process. For example, the condition of a main crossing beneath a major highway may be unknown. But the consequence of failure may be quite high. If the main has been in service past its expected lifespan and is likely to fail, the risk of failure is considered high enough to begin a condition assessment of this asset. Knowledge of pipe materials and failure histories, geological and geographical environment, and other data will help develop a credible likelihood of failure rating for each asset or group of assets. The following are five steps to successfully integrate asset risk management into the capital financing process.

Step 1—Establish the Consequence of Failure

After developing an inventory of utility assets, estimate the consequences of failure for each asset or group of assets. For this step, divide the water system into several broad consequence categories. Within each consequence category add one or more subcategories to ensure that all of the consequence factors for the broad category are included. Next, define the consequence of failure and classify them into ratings based on severity. This process is more subjective and usually results in establishing one to five or six levels of severity. The City of Portland, Ore., uses five levels of severity—from 1 being very low to 5 indicating extreme severity. Table 1-3, the risk assessment matrix in chapter 1, also uses five levels of severity for the consequences of failure. Table A-2 at the end of this appendix illustrates how the City of Portland divided the consequences of failure into seven broad categories and then into 26 subcategories, with tiers or rates of severity allocated to each subcategory. For example, the city defined severe (Tier 5) consequences for a "Service Outage" to be a major or large service area out of water for longer than 72 hr, or half the city without water for 8 hr or more, while a moderate (Tier 3) consequence is 250 to 1,000 services out of water for 24 to 72 hr In the "Health and safety" category, a Tier 5 consequence is a single loss of life, and a Tier 3 consequence would be a major injury with greater than 30 days off work. In the category of "Community financial impact" the Tier 1 consequence is more than $20 million in losses, while the Tier 3 impact is more than $5 million in losses.

Not all assets will impact all consequence categories with the same degree of severity. If an asset falls into more than one severity rating, the conservative approach is to match the asset to the level of highest severity among all consequence categories and/or subcategories and use that value to plot the asset's consequence of failure on the risk matrix. Eventually, each asset or group of assets should end up with a single consequence of failure rating of 1 to 5.

Step 2—Estimate the Likelihood of Failure

The likelihood of failure can be estimated using intervals between major weather or seismic events, and/or asset conditions resulting in an observable failure frequency. By identifying event timeframes in a table, a utility can differentiate between an imminent failure and a failure likely to be caused by a severe weather event, an earthquake, or a landslide (Table A-1). Other indicators of failure frequency can be used as well, depending on local circumstances.

The risk assessment matrix in Table 1-3 in chapter 1 uses a scale of 1 to 5, but a longer scale could also be used. It should be noted that too many scales or levels only adds complexity without increasing consistency in the risk assessment process.

The key to a likelihood-of-failure table is clear and consistent assignment of failure rates that are close to actual experience. Failure rates and their causes may not be known in the beginning. However, they can be refined and adjusted as more failure history and experience is accumulated.

Table A-1 Likelihood of a failure based on potential for a severe weather event, earthquake, fire, or landslide

Risk Rating		Frequency of Failure
5	Likely	May occur once a year
4	Possible	May occur once in 2 to 10 years
3	Unlikely	May occur once in 11 to 50 years
2	Very unlikely	May occur once in 51 to 100 years
1	Remotely possible	Occurs less than once in 100+ years

Step 3—Create an Asset Risk Score and/or Matrix

After developing consequence of failure and likelihood of failure rates, utilities can create an *asset risk score* for each asset or group of assets by multiplying the consequence of failure rate by the likelihood of failure rate. These asset risk scores can either be plotted on an asset risk matrix or just listed in numerical order. Multiplying the two failure rates may yield two different assets with the same combined score. For example, a failure rate of 5, times a consequence rate of 1, yields an asset risk score of 5. If the failure rate of another asset is 1 and the consequence rate is 5, the asset risk score is also 5. As utilities develop asset risk scores and a risk matrix, it is important to clearly define the asset risk scores so that assets with the same combined risk score are consistently classified as an extremely low, low, medium, medium high, high, or extremely high risk to the utility.

Plotting asset risk scores on the risk matrix creates a profile of the water system's vulnerability to consequential asset failures. The utility now has a defensible argument for focusing capital improvement plans and dollars on replacing or rebuilding the system's higher risk assets. Which quadrant or cell the assets fall in the matrix will determine the degree of risk mitigation that the utility should perform.

Step 4—Develop a Risk-Related Service Level

The next step is to integrate risks into a *risk-related service level*. For example, Portland's risk-related service level requires addressing extreme risks within the next one to three years. If a risk is high, it is to be addressed within five years. The risk-related service level is important because mitigating the risk of asset failure can be a driver for capital investment decisions.

Step 5—Apply Risk Mitigation to Asset Management

The last step is to tie asset risk management into the capital investment process. Incorporating risk service levels into the capital financial planning process forces the utility to answer whether it can justify the investments needed to mitigate risk and whether benefits will outweigh costs. Cost–benefit analysis can be helpful in answering these questions. In some cases benefits will outweigh costs, in others it will not. But at least risk mitigation has been discussed and considered in the development of capital and financial plans and budgets.

Table A-2 Consequence of failure

	Tier 5	Tier 4	Tier 3	Tier 2	Tier 1
Severity Rating	Extreme	High	Moderate	Low	Very Low
Regulations					
MCL violation	Revocation of public water system certification	Tier 1 violation (acute), e.g., consent order to treat	Tier 2 Notification (includes nitrates, etc.) Secondary MCL problem, e.g., manganese problems	Monitoring/ reporting requirement Discolored water	Potential news report on water quality issue
Boil water		Boil-water order or other public warning to half of city or more	Boil-water order to smaller service area		
Treatment violation		Event triggers loss of filtration exemption	Sampling violations	Less than 100% compliance with state and federal water quality regulations	Detects requiring notification
Supply					
Outages	Major or large service area out of water for longer than 72 hr Half the city out of water for 8 hr or more	Major or large service area out of water for 24–72 hr 250–1,000 service,s out of water for longer than 72 hr Supply lines out of service for longer than 72 hr	250–1,000 services out of water for 24–72 hr Supply lines out of service for 24–72 hr 50–250 services out of water for longer than 72 hr Major or large service area out of water for 8 hr	50–250 services out of water for 24–72 hr Up to 50 services out of water for > 72 hr 250–1,000 services out of water for 8 hr or more	Customer out of water more than 3 times per year Up to 250 services out of water for 8 hr or more
Low service pressures			Half or more of city receiving pressure < 20 psi for 2–4 hr Major or large service area receiving pressure < 20 psi for 72 hr or more	Major or large service area receiving pressure < 20 psi for 2–4 hr or more Up to 250 services receiving pressure < 20 psi for 72 hr or more	50–250 services receiving pressure < 20 psi for 2–4 hr or more
Excessive pressure			250 services provided water at higher of 150 psi or 50 psi over normal for 1 hr	50 services provided water at higher of 150 psi or 50 psi over normal for 1 hr	

(Table continued on next page)

Table A-2 Consequence of failure (continued)

	Tier 5	Tier 4	Tier 3	Tier 2	Tier 1
Severity Rating	Extreme	High	Moderate	Low	Very Low
Fire flows			Supplying less than minimal fire flow to a commercial/ industrial fire	Supplying less than minimal fire flow to a residential fire Supplying minimal fire flow to a commercial/ industrial fire	Supplying less than reduced fire flow to a residential fire Supplying reduced fire flow to a commercial/ industrial fire
Environmental					
Chlorinated water release	Release of a volume of chlorinated water that results in a major fish kill of ESA-listed species*	Release of chlorinated water that results in visible die off of aquatic life in a river or stream	Release of chlorinated water that results in moderate damage to aquatic habitat	Release of chlorinated water that results in minor damage to aquatic habitat	
Erosion		Erosion causes major damage to sensitive aquatic or terrestrial habitats	Erosion causes moderate damage to sensitive aquatic or terrestrial habitats	Erosion causes minor damage to sensitive aquatic or terrestrial habitats	Erosion causes mud to track into street and silt up catch basins
Conservation and sustainability			Increase in utility's carbon emissions from 2007 level or significant increase in per capita water use in retail service area	Decrease in energy generated by renewable sources	
Energy use			Increase energy use significantly	Increase energy use moderately	Increase energy use
Loss of revenue					
Weather impact on water supply		Water shortage—mandatory curtailment	Voluntary curtailment	Strong conservation message	Conservation message
Wholesale customer agreement		Bureau does not deliver guarantee > 30 days twice in 10 years—wholesalers can reduce future demand 10%	Customer dispute or litigation more than $5 million OR customers with 10-year agreement do not renew	Bureau does not deliver guarantee > 5 days—wholesalers does not have to pay	
Billing and collection issues	Bills not generated (or $$ not collected) for 60 days	Bills not generated (or $$ not collected) for 30 days	Bills not generated (or $$ not collected) for 15 days		

(Table continued on next page)

Table A-2 Consequence of failure (continued)

	Tier 5	Tier 4	Tier 3	Tier 2	Tier 1
Severity Rating	Extreme	High	Moderate	Low	Very Low
Large expenditures					
Bureau financial impact	Additional expenditures/ loss > $20 million	Additional expenditures/ loss > $10 million	Additional expenditures/ loss > $5 million	Additional expenditures/ loss > $1 million	
Bond-rating downgrade	Moody's Rating C		Moody's Rating B	Moody's Rating A	
Liability insurance coverage thresholds	> $30 million in damages	$10–30 million in damages	$1–10 million in damages	$0.5–1 million in damages	
Public confidence					
Use restrictions		Mandatory curtailment measures	Request for voluntary reduction in water use; interruption to supply to wholesalers		
Security breach	Successful/proven contamination action		News report on contamination attempt		
Customer billing		Equity: 30% or more of bills/ accounts not paying their fair share	Equity: 10% or more of bills/ accounts not paying their fair share		
		Accuracy: miscalculation of large number of bills, 25% error	Accuracy: miscalculation of large number of bills, 10% error		
Customer response	No contact or unable to respond to customers adequately, 30 days	No contact or unable to respond to customers adequately, 20 days	No contact or unable to respond to customers adequately, 7 days	No contact or unable to respond to customers adequately, 5 days	Unable to answer 80% of calls within 60 seconds

(Table continued on next page)

Table A-2 Consequence of failure (continued)

	Tier 5	Tier 4	Tier 3	Tier 2	Tier 1
Severity Rating	Extreme	High	Moderate	Low	Very Low
Customer satisfaction			< 50% give a High or Very High rating on auditor's service efforts and accomplishments (SEA) survey	> 7 complaints per 1,000 customers per year	< 75% give a High or Very High rating on auditor's SEA survey
Social consequences					
Health and safety	Single loss of life	Permanent disability	Major injury, > 30 days off work Life-safety issues	Work time injury	Inconvenience only
Water quality	Large scale water system contamination, causing a single fatality	Widespread sickness	Water quality exceedance Minor quality issues, such as turbidity event	Localized water quality issues such as turbidity from main break, etc.	Localized water quality issue (nonhealth): color, taste, localized turbidity problem
Illness		> 100 hospitalized due to confirmed waterborne disease	> 10 confirmed cases of waterborne disease Unconfirmed reports of minor illness with symptoms related to waterborne disease		
Community financial impact	> $20 million of losses	> $10 million of losses	> $5 million of losses	> $1 million of losses	

*Endangered Species Act
Source: City of Portland, Ore.

This page intentionally blank.

Appendix B

Comparison of Capital Financing Scenarios

No single correct capital financing strategy is applicable to all water utilities in all situations. Each utility must select a strategy specifically tailored to meet its own unique financial, operational, regulatory, and political challenges. Although there is no one-size-fits-all approach to capital financing, it is possible to illustrate the financial outcomes created by the use of varying levels of debt financing. This appendix compares the forecasted financial results, over a 20-year period, for a hypothetical water utility under three capital funding strategies featuring varying degrees of debt financing:

- Strategy 1: 0% use of debt financing
- Strategy 2: Balanced use of debt financing and rate revenue funding (40% debt)
- Strategy 3: Heavy reliance on debt financing (80% debt)

All three scenarios feature the exact same level of forecasted capital improvement program expenditures, operating and maintenance expenditures, and billed water consumption. The annual water rate increases specified in each scenario have been set to achieve a positive annual operating surplus and a year-end operating reserve equivalent to at least 90 days of annual operating and maintenance expenditures. This level of operating reserves is often targeted by water utilities to help cope with climatically induced fluctuations in water rate revenues and/or unexpected operating contingencies. The annual rate increases specified in each scenario have also been set to maintain a minimum 1.25× debt-service coverage ratio where applicable. This level of debt-service coverage is commonly required by investors who purchase municipal utility debt. Finally, each scenario also assumes that the hypothetical utility will not lower water rates during those years when the forecasted water rate revenues exceed forecasted expenditures. This allows for the accumulation of cash reserves that can be used to offset future capital expenditures. Table B-1 of key inputs provides a summary of the assumptions used in each capital funding scenario.

Table B-1 Key inputs for summary outcome tables (B-2, B-3, and B-4)

Base Year Billed-Water Consumption (gal)	
Customer Class	Base Year
Single-family residential	1,500,000
Multifamily residential	900,000
Commercial	1,200,000
Irrigation	600,000
Total billed consumption	4,200,000
Growth in Billed Water Consumption	
Annual growth in customer accounts	1.00%
Annual change in average consumption per account	0.00%
Debt Financing	
Senior lien interest rate	5.00%
Senior lien financing term	20 years
Minimum required combined debt-service coverage ratio	1.20×
Operating Expenditures	
Annual cost escalation / Inflation rate	3.00%

STRATEGY 1: 0% DEBT FINANCING

Under this strategy, all capital expenditures are funded on a pay-as-you-go basis via current year water rate revenues. This approach is used by utilities that, for whatever reason, are unwilling or unable to access the capital markets to obtain debt financing. Under this capital financing strategy, the entire cost of constructing capital assets must be paid by current ratepayers. The potential advantages of 0% debt financing include:

- Large increase in cash reserves during periods of reduced or stable capital expenditures
- The avoidance of debt financing implies lower long-term financial risk and greater long-term financial flexibility

Potential disadvantages are:

- Large rate increases are required during periods of heavy capital expenditures
- Current ratepayers pay 100% of the cost of capital

Table B-2, Summary outcomes for Strategy 1: 0% Debt Financing, which follows this narrative, shows that the decision to reject the use of any debt financing results in the need for large annual water rate increases in Years 1–8 (Line 3). This causes water rates to increase by approximately 200% from $2.90 per thousand gallons in Year 1 to $5.88 per thousand gallons in Year 8 (Line 4). These large rate increases are required to fund forecasted capital expenditures while maintaining a positive operating surplus and a year-end operating reserve equivalent to at least 90 days of annual operating expenditures (Lines 10 and 11).

Beginning in Year 9, forecast capital expenditures fall from $19 million to only $15 million annually (Line 17). From this point forward, the utility begins to accumulate large year-end cash reserves (Line 11) without the need to increase water rates (Line 3). Because no water rate increases are required, the utility can continue to charge only $5.88 per thousand gallons throughout the remainder of the 20-year planning horizon in Years 9–20. Finally, as shown in Lines 23–30, the utility has no annual debt service payments (Line 27), no debt-service coverage requirements (Line 28), and no outstanding debt (Line 30). Thus, the large rate increases imposed on current customers in Years 1–8 resulted in the utility achieving a very strong financial position and a high level of financial flexibility without the need to impose any rate increases in Years 9–20. It is important to note that for many utilities, the outcomes shown in Years 9–20 may not be acceptable to either utility ratepayers or governing bodies. Specifically, the accumulation of such large operating reserves of the type shown in Line 11 would be considered excessive absent a specific justification for accumulating such a large amount of ratepayer-funded cash.

STRATEGY 2: BALANCED USE OF DEBT FINANCING AND RATE REVENUE FUNDING (40% DEBT)

This strategy provides a relative balance between the use of debt financing and rate revenue funding. The example shown assumes that debt financing is used to fund 40% of annual capital expenditures while current year rate revenues fund 60%. Regardless of the specific combination debt and rate revenue used (i.e., 40% debt, 50% debt, etc.), this balanced approach is often used by water utilities to minimize the water rate increases imposed on current ratepayers. The potential advantages of this approach include:

- The need for large rate increases during periods of heavy capital expenditures is reduced
- Current ratepayers do not need to pay the full 100% of the cost of capital assets that will also benefit future ratepayers

Disadvantages include:

- The use of debt financing increases overall utility costs (interest payments and bond issuance costs)
- The use of debt financing implies marginally higher long-term financial risk and lower long-term financial flexibility

Table B-3, Summary outcomes for Strategy 2: Balanced use of debt financing and rate revenue funding (40% Debt), shows that the decision to fund 40% of capital expenditures using debt allows the utility to avoid any rate increases in Years 1 and 2 (Line 3). As a result, water rates stay at their existing level of only $2.60 per thousand gallons during these initial two years of the 20-year planning horizon (Line 4). The use of 40% debt financing to fund capital expenditures also has a positive long-term impact on water rates. For example, water rates in Year 9 are forecast to be only $4.89 per thousand gallons, compared with $5.88 per thousand gallons in Strategy 1 (0% Debt). In addition, the rates forecast to be charged in Year 20 are only $5.65 per thousand gallons, compared with $5.88 under Strategy 1. This demonstrates that the use of debt to fund capital expenditures can help to offset the need to impose large rate increases on customers in order to fund capital expenditures.

Despite the positive impacts for utility ratepayers, the use of debt financing does have some potential drawbacks. For example, the utility will not accumulate large year-end operating reserves equal to those under Strategy 1 during the latter years of the planning horizon. As shown in Line 11, the forecast Year 20 operating reserve under Strategy 2 is approximately $8.2 million, compared with $51.8 million under Strategy 1. This lower operating reserve occurs because the utility must make the annual debt service payments that rise to $8.9 million annually in Year 20 (Line 27). Further, the utility is forecast to have outstanding debt principal of $71.7 million in Year 20 compared with $0 outstanding debt under Strategy 1. These outcomes clearly demonstrate that benefits of lower rate increases created by the moderate use of debt are, to at least some degree, accompanied by a slight reduction in financial strength and financial flexibility when measured by the amount of operating reserves and the degree of financial leverage in a utility's capital structure.

STRATEGY 3: HEAVY RELIANCE ON DEBT FINANCING (80% DEBT)

Under this strategy, a utility chooses to pay for a large majority of its capital expenditures via the use of debt financing. In the example shown, debt financing is assumed to fund 80% of annual capital expenditures. This heavy reliance on financial leverage is not a long-term strategy used by most water utilities. Nonetheless, it is useful to review the financial outcomes associated with this strategy because they contrast so sharply with more normative capital financing approaches. One potential advantage of this approach is that large rate increases are mitigated, *but only in the short term.*

Potential disadvantages are greater and include:

- Heavy use of debt financing results in higher rate increases over the long term because of the need to comply with debt-service coverage requirements
- Heavy use of financing greatly increases overall utility costs (interest payments and bond issuance costs)
- Heavy use of debt financing significantly increases financial risk and significantly lowers long-term financial flexibility because of the high percentage of revenues that must be devoted to debt service

Table B-4, Summary Outcomes for Strategy 3: Heavy Reliance on Debt Financing (80% Debt), shows that the decision to fund 80% of capital expenditures using debt has a very positive impact on water rates during the first-half of the 20-year planning horizon. For example, in Year 10, the utility need charge only $4.16 per thousand gallons (Line 4) compared with $5.88 per thousand gallons in Strategy 1 (0% Debt) and $4.89 per thousand gallons in Strategy 2 (40% Debt).

Unfortunately, this positive rate outcome in the first-half of the planning horizon comes with a heavy price by Year 20. With every dollar of capital expenditures being funded 80% by debt, the utility accumulates total outstanding debt of $143.3 million in Year 20 (Line 30) compared with $0 under Strategy 1 and $71.7 million under Strategy 2. The annual debt service forecast for Year 20 grows to $17.8 million (Line 27) compared with $0 under Strategy 1 and $8.9 million under Strategy 2. Thus, by Year 20, approximately 54% of annual operating revenues are devoted to paying debt service and maintaining the minimum 1.25× debt-service coverage ratio ($17.8 million debt service from Line 27 divided by $33.2 million in water rate revenues from Line 6) Many utility financial managers, credit ratings agencies, and investors would consider this high degree of financial leverage to significantly reduce the utility's flexibility to respond to unforeseen operating contingencies or make additional large capital expenditures beyond Year 20 without the imposition of massive rate increases.

Further, in order to maintain the minimum required 1.25× debt-service coverage ratio, the utility must impose large rate increases in Years 11–20. The forecast Year 20 rate of $6.50 per thousand gallons (Line 4) is approximately 17% higher than the rate of $5.88 per thousand gallons required under Strategy 1 and 41% higher than the rate of $4.89 per thousand gallons required under Strategy 2. These outcomes clearly demonstrate that an excessive reliance on debt financing to fund forecast capital expenditures can result in very negative outcomes when viewed from a long-term perspective.

The graphs depicted in Figures B-1 through B-7 provide a visual representation of key financial metrics associated with each capital funding strategy discussed here.

Table B-2 Summary outcomes for Strategy 1: 0% Debt Financing

Line No.	Operating Performance	Forecast Year 1	Forecast Year 2	Forecast Year 3	Forecast Year 4	Forecast Year 5	Forecast Year 6	Forecast Year 7	Forecast Year 8	Forecast Year 9	Forecast Year 10	Forecast Year 15	Forecast Year 20
1	Billed consumption (1000s of gal)	4,242,700	4,285,522	4,328,468	4,371,538	4,414,853	4,459,607	4,504,608	4,550,475	4,596,466	4,642,579	4,875,857	5,113,198
2	Water rates												
3	Annual rate revenue increases	11.75%	8.50%	15.00%	13.00%	5.50%	15.50%	13.50%	3.75%	0.00%	0.00%	0.00%	0.00%
4	Average rate per 1,000 gal	$2.90	$3.15	$3.62	$4.10	$4.32	$4.99	$5.66	$5.88	$5.88	$5.88	$5.88	$5.88
5	Operating surplus/deficit												
6	Rate revenues	$12,323,107	$13,505,873	$15,687,836	$17,904,397	$19,077,219	$22,256,464	$25,515,003	$26,739,353	$27,008,252	$27,278,248	$28,646,074	$30,044,115
7	Less: O&M expenditures	($6,249,062)	($6,436,533)	($6,629,629)	($6,828,518)	($7,033,374)	($7,244,375)	($7,461,706)	($7,685,557)	($7,916,124)	($8,153,608)	($9,452,266)	($10,957,767)
8	Less: Debt service	$0	$0	$0	$0	$0	$0	$0	$0	$0	$0	$0	$0
9	Less: Rate revenue funding of CIP	($6,000,000)	($7,000,000)	($9,000,000)	($11,000,000)	($12,000,000)	($15,000,000)	($18,000,000)	($19,000,000)	($15,000,000)	($15,000,000)	($15,000,000)	($15,000,000)
10	Net operating surplus/deficit	$74,045	$69,339	$58,207	$75,879	$43,845	$12,089	$53,297	$53,796	$4,092,128	$4,124,640	$4,193,807	$4,086,348
11	Operating reserve ending balance	$1,574,045	$1,643,385	$1,701,592	$1,777,471	$1,821,316	$1,833,405	$1,886,702	$1,940,497	$6,032,625	$10,157,266	$31,052,446	$51,776,009
12	Operating reserve target (90 days O&M)	($1,562,265)	($1,609,133)	($1,657,407)	($1,707,130)	($1,758,343)	($1,811,094)	($1,865,427)	($1,921,389)	($1,979,031)	($2,038,402)	($2,363,067)	($2,739,442)

Line No.	Capital Financing Strategy		Forecast Year 1	Forecast Year 2	Forecast Year 3	Forecast Year 4	Forecast Year 5	Forecast Year 6	Forecast Year 7	Forecast Year 8	Forecast Year 9	Forecast Year 10	Forecast Year 15	Forecast Year 20
13	CIP expenditures													
14	Projects – Long-term growth		$0	$0	$0	$0	$2,000,000	$2,000,000	$3,000,000	$2,000,000	$1,000,000	$0	$0	$0
15	Projects – Regulatory mandates		$0	$0	$2,000,000	$4,000,000	$3,000,000	$4,000,000	$4,000,000	$5,000,000	$2,000,000	$0	$0	$0
16	Projects – Renewal of existing assets		$6,000,000	$7,000,000	$7,000,000	$7,000,000	$7,000,000	$9,000,000	$11,000,000	$12,000,000	$12,000,000	$15,000,000	$15,000,000	$15,000,000
17	Total CIP expenditures		$6,000,000	$7,000,000	$9,000,000	$11,000,000	$12,000,000	$15,000,000	$18,000,000	$19,000,000	$15,000,000	$15,000,000	$15,000,000	$15,000,000
18	CIP financing sources													
19	Rate revenue funding	100%	$6,000,000	$7,000,000	$9,000,000	$11,000,000	$12,000,000	$15,000,000	$18,000,000	$19,000,000	$15,000,000	$15,000,000	$15,000,000	$15,000,000
20	Senior lien debt funding	0%	$0	$0	$0	$0	$0	$0	$0	$0	$0	$0	$0	$0
22	Total sources of financing	100%	$6,000,000	$7,000,000	$9,000,000	$11,000,000	$12,000,000	$15,000,000	$18,000,000	$19,000,000	$15,000,000	$15,000,000	$15,000,000	$15,000,000

Line No.	Debt Metrics	Forecast Year 1	Forecast Year 2	Forecast Year 3	Forecast Year 4	Forecast Year 5	Forecast Year 6	Forecast Year 7	Forecast Year 8	Forecast Year 9	Forecast Year 10	Forecast Year 15	Forecast Year 20
23	Debt-service coverage												
24	Revenue	$12,323,107	$13,505,873	$15,687,836	$17,904,397	$19,077,219	$22,256,464	$25,515,003	$26,739,353	$27,008,252	$27,278,248	$28,646,074	$30,044,115
25	Less: O&M expenditures	($6,249,062)	($6,436,533)	($6,629,629)	($6,828,518)	($7,033,374)	($7,244,375)	($7,461,706)	($7,685,557)	($7,916,124)	($8,153,608)	($9,452,266)	($10,957,767)
26	Net revenues available for debt service	$6,074,045	$7,069,339	$9,058,207	$11,075,879	$12,043,845	$15,012,089	$18,053,297	$19,053,796	$19,092,128	$19,124,640	$19,193,807	$19,086,348
27	Senior lien debt service	$0	$0	$0	$0	$0	$0	$0	$0	$0	$0	$0	$0
28	Debt-service coverage ratio	N/A	N/A	N/A	N/A	N/A	N/A	N/A	N/A	N/A	N/A	N/A	N/A
29	Outstanding debt principal												
30	Year-end senior lien debt principal	$0	$0	$0	$0	$0	$0	$0	$0	$0	$0	$0	$0
31	Total outstanding debt	$0	$0	$0	$0	$0	$0	$0	$0	$0	$0	$0	$0

Table B-3 Summary outcomes for Strategy 2: 40% Debt Financing

Line No.	Operating Performance		Forecast Year 1	Forecast Year 2	Forecast Year 3	Forecast Year 4	Forecast Year 5	Forecast Year 6	Forecast Year 7	Forecast Year 8	Forecast Year 9	Forecast Year 10	Forecast Year 15	Forecast Year 20
1	Billed consumption (1000s of gal)		4,242,700	4,285,522	4,328,468	4,371,538	4,414,853	4,459,607	4,504,608	4,550,475	4,596,466	4,642,579	4,875,857	5,113,198
2	Water rates													
3	Annual rate revenue increases		0.00%	0.00%	13.50%	12.50%	7.25%	14.75%	13.25%	5.75%	0.00%	0.00%	2.00%	1.75%
4	Average rate per 1,000 gal		$2.60	$2.60	$2.95	$3.32	$3.56	$4.08	$4.63	$4.89	$4.89	$4.89	$5.13	$5.65
5	Operating surplus/deficit													
6	Rate revenues		$11,027,407	$11,138,973	$12,769,836	$14,509,497	$15,716,419	$18,216,664	$20,837,703	$22,258,653	$22,482,152	$22,706,748	$24,994,374	$28,871,815
7	Less: O&M expenditures		($6,249,062)	($6,436,533)	($6,629,629)	($6,828,518)	($7,033,374)	($7,244,375)	($7,461,706)	($7,685,557)	($7,916,124)	($8,153,608)	($9,452,266)	($10,957,767)
8	Less: Debt service		($192,582)	($417,261)	($706,135)	($1,059,202)	($1,444,367)	($1,925,822)	($2,503,569)	($3,113,412)	($3,594,868)	($4,076,323)	($6,483,601)	($8,890,879)
9	Less: Rate revenue funding of CIP		($3,600,000)	($4,200,000)	($5,400,000)	($6,600,000)	($7,200,000)	($9,000,000)	($10,800,000)	($11,400,000)	($9,000,000)	($9,000,000)	($9,000,000)	($9,000,000)
10	Net operating surplus/deficit		$985,763	$85,178	$34,072	$21,777	$38,679	$46,467	$72,428	$59,683	$1,971,160	$1,476,817	$58,506	$23,169
11	Operating reserve ending balance		$2,485,763	$2,570,941	$2,605,013	$2,626,790	$2,665,469	$2,711,936	$2,784,364	$2,844,047	$4,815,207	$6,292,024	$7,960,254	$8,158,279
12	Operating reserve target (90 days O&M)		($1,562,265)	($1,609,133)	($1,657,407)	($1,707,130)	($1,758,343)	($1,811,094)	($1,865,427)	($1,921,389)	($1,979,031)	($2,038,402)	($2,363,067)	($2,739,442)
Line No.	**Capital Financing Strategy**		**Forecast Year 1**	**Forecast Year 2**	**Forecast Year 3**	**Forecast Year 4**	**Forecast Year 5**	**Forecast Year 6**	**Forecast Year 7**	**Forecast Year 8**	**Forecast Year 9**	**Forecast Year 10**	**Forecast Year 15**	**Forecast Year 20**
13	CIP expenditures													
14	Projects – Long-term growth		$0	$0	$0	$0	$2,000,000	$2,000,000	$3,000,000	$2,000,000	$1,000,000	$0	$0	$0
15	Projects – Regulatory mandates		$0	$0	$2,000,000	$4,000,000	$3,000,000	$4,000,000	$4,000,000	$5,000,000	$2,000,000	$0	$0	$0
16	Projects – Renewal of existing assets		$6,000,000	$7,000,000	$7,000,000	$7,000,000	$7,000,000	$9,000,000	$11,000,000	$12,000,000	$12,000,000	$15,000,000	$15,000,000	$15,000,000
17	Total CIP expenditures		$6,000,000	$7,000,000	$9,000,000	$11,000,000	$12,000,000	$15,000,000	$18,000,000	$19,000,000	$15,000,000	$15,000,000	$15,000,000	$15,000,000
18	CIP financing sources													
19	Rate revenue funding	60%	$4,200,000	$5,400,000	$6,600,000	$7,200,000	$9,000,000	$10,800,000	$11,400,000	$9,000,000	$9,000,000	$9,000,000	$9,000,000	$15,000,000
20	Senior lien debt funding	40%	$2,800,000	$3,600,000	$4,400,000	$4,800,000	$6,000,000	$7,200,000	$7,600,000	$6,000,000	$6,000,000	$6,000,000	$6,000,000	$0
22	Total sources of financing	100%	$7,000,000	$9,000,000	$11,000,000	$12,000,000	$15,000,000	$18,000,000	$19,000,000	$15,000,000	$15,000,000	$15,000,000	$15,000,000	$15,000,000
Line No.	**Debt Metrics**		**Forecast Year 1**	**Forecast Year 2**	**Forecast Year 3**	**Forecast Year 4**	**Forecast Year 5**	**Forecast Year 6**	**Forecast Year 7**	**Forecast Year 8**	**Forecast Year 9**	**Forecast Year 10**	**Forecast Year 15**	**Forecast Year 20**
23	Debt-service coverage													
24	Revenue		$11,027,407	$11,138,973	$12,769,836	$14,509,497	$15,716,419	$18,216,664	$20,837,703	$22,258,653	$22,482,152	$22,706,748	$24,994,374	$28,871,815
25	Less: O&M expenditures		($6,249,062)	($6,436,533)	($6,629,629)	($6,828,518)	($7,033,374)	($7,244,375)	($7,461,706)	($7,685,557)	($7,916,124)	($8,153,608)	($9,452,266)	($10,957,767)
26	Net revenues available for debt service		$4,778,345	$4,702,439	$6,140,207	$7,680,979	$8,683,045	$10,972,289	$13,375,997	$14,573,096	$14,566,028	$14,553,140	$15,542,107	$17,914,048
27	Senior lien debt service		$192,582	$417,261	$706,135	$1,059,202	$1,444,367	$1,925,822	$2,503,569	$3,113,412	$3,594,868	$4,076,323	$6,483,601	$8,890,879
28	Debt-service coverage ratio		24.81	11.27	8.70	7.25	6.01	5.70	5.34	4.68	4.05	3.57	2.40	2.01
29	Outstanding debt principal													
30	Year-end senior lien debt principal		$2,400,000	$5,127,418	$8,566,527	$12,688,719	$17,063,953	$22,465,525	$28,855,721	$35,387,681	$40,036,394	$44,436,088	$62,241,019	$71,663,396
31	Total outstanding debt		$2,400,000	$5,127,418	$8,566,527	$12,688,719	$17,063,953	$22,465,525	$28,855,721	$35,387,681	$40,036,394	$44,436,088	$62,241,019	$71,663,396

Table B-4 Summary outcomes for Strategy 3: 80% Debt Financing

Line No.	Operating Performance		Forecast Year 1	Forecast Year 2	Forecast Year 3	Forecast Year 4	Forecast Year 5	Forecast Year 6	Forecast Year 7	Forecast Year 8	Forecast Year 9	Forecast Year 10	Forecast Year 15	Forecast Year 20
1	Billed consumption (1000s of gal)		4,242,700	4,285,522	4,328,468	4,371,538	4,414,853	4,459,607	4,504,608	4,550,475	4,596,466	4,642,579	4,875,857	5,113,198
2	Water rates													
3	Annual rate revenue increases		0.00%	0.00%	0.00%	0.00%	7.50%	13.50%	13.00%	9.00%	2.50%	4.00%	4.50%	3.75%
4	Average rate per 1,000 gal		$2.60	$2.60	$2.60	$2.60	$2.79	$3.17	$3.58	$3.91	$4.00	$4.16	$5.24	$6.50
5	Operating surplus/deficit													
6	Rate revenues		$11,027,407	$11,138,973	$11,250,936	$11,363,297	$12,337,219	$14,143,964	$16,143,403	$17,774,153	$18,401,652	$19,328,948	$25,535,674	$33,212,115
7	Less: O&M expenditures		($6,249,062)	($6,436,533)	($6,629,629)	($6,828,518)	($7,033,374)	($7,244,375)	($7,461,706)	($7,685,557)	($7,916,124)	($8,153,608)	($9,452,266)	($10,957,767)
8	Less: Debt service		($385,164)	($834,523)	($1,412,270)	($2,118,404)	($2,888,733)	($3,851,644)	($5,007,137)	($6,226,825)	($7,189,736)	($8,152,647)	($12,967,202)	($17,781,757)
9	Less: Rate revenue funding of CIP		($1,200,000)	($1,400,000)	($1,800,000)	($2,200,000)	($2,400,000)	($3,000,000)	($3,600,000)	($3,800,000)	($3,000,000)	($3,000,000)	($3,000,000)	($3,000,000)
10	Net operating surplus/deficit		$3,193,181	$2,467,917	$1,409,037	$216,375	$15,112	$47,945	$74,559	$61,771	$295,792	$22,693	$116,205	$1,472,591
11	Operating reserve ending balance		$4,693,181	$7,161,098	$8,570,135	$8,786,509	$8,801,622	$8,849,566	$8,924,126	$8,985,897	$9,281,689	$9,304,382	$9,514,563	$14,635,749
12	Operating reserve target (90 days O&M)		($1,562,265)	($1,609,133)	($1,657,407)	($1,707,130)	($1,758,343)	($1,811,094)	($1,865,427)	($1,921,389)	($1,979,031)	($2,038,402)	($2,363,067)	($2,739,442)
Line No.	**Capital Financing Strategy**		**Forecast Year 1**	**Forecast Year 2**	**Forecast Year 3**	**Forecast Year 4**	**Forecast Year 5**	**Forecast Year 6**	**Forecast Year 7**	**Forecast Year 8**	**Forecast Year 9**	**Forecast Year 10**	**Forecast Year 15**	**Forecast Year 20**
13	CIP expenditures													
14	Projects – Long-term growth		$0	$0	$0	$0	$2,000,000	$2,000,000	$3,000,000	$2,000,000	$1,000,000	$0	$0	$0
15	Projects – Regulatory mandates		$0	$0	$2,000,000	$4,000,000	$3,000,000	$4,000,000	$4,000,000	$5,000,000	$2,000,000	$0	$0	$0
16	Projects – Renewal of existing assets		$6,000,000	$7,000,000	$7,000,000	$7,000,000	$7,000,000	$9,000,000	$11,000,000	$12,000,000	$12,000,000	$15,000,000	$15,000,000	$15,000,000
17	Total CIP expenditures		$6,000,000	$7,000,000	$9,000,000	$11,000,000	$12,000,000	$15,000,000	$18,000,000	$19,000,000	$15,000,000	$15,000,000	$15,000,000	$15,000,000
18	CIP financing sources													
19	Rate revenue funding	20%	$1,400,000	$1,800,000	$2,200,000	$2,400,000	$3,000,000	$3,600,000	$3,800,000	$3,000,000	$3,000,000	$3,000,000	$3,000,000	$15,000,000
20	Senior lien debt funding	80%	$5,600,000	$7,200,000	$8,800,000	$9,600,000	$12,000,000	$14,400,000	$15,200,000	$12,000,000	$12,000,000	$12,000,000	$12,000,000	$0
22	Total sources of financing	100%	$7,000,000	$9,000,000	$11,000,000	$12,000,000	$15,000,000	$18,000,000	$19,000,000	$15,000,000	$15,000,000	$15,000,000	$15,000,000	$15,000,000
Line No.	**Debt Metrics**		**Forecast Year 1**	**Forecast Year 2**	**Forecast Year 3**	**Forecast Year 4**	**Forecast Year 5**	**Forecast Year 6**	**Forecast Year 7**	**Forecast Year 8**	**Forecast Year 9**	**Forecast Year 10**	**Forecast Year 15**	**Forecast Year 20**
23	Debt-service coverage													
24	Revenue		$11,027,407	$11,138,973	$11,250,936	$11,363,297	$12,337,219	$14,143,964	$16,143,403	$17,774,153	$18,401,652	$19,328,948	$25,535,674	$33,212,115
25	Less: O&M expenditures		($6,249,062)	($6,436,533)	($6,629,629)	($6,828,518)	($7,033,374)	($7,244,375)	($7,461,706)	($7,685,557)	($7,916,124)	($8,153,608)	($9,452,266)	($10,957,767)
26	Net revenues available for debt service		$4,778,345	$4,702,439	$4,621,307	$4,534,779	$5,303,845	$6,899,589	$8,681,697	$10,088,596	$10,485,528	$11,175,340	$16,083,407	$22,254,348
27	Senior lien debt service		$385,164	$834,523	$1,412,270	$2,118,404	$2,888,733	$3,851,644	$5,007,137	$6,226,825	$7,189,736	$8,152,647	$12,967,202	$17,781,757
28	Debt-service coverage ratio		12.41	5.63	3.27	2.14	1.84	1.79	1.73	1.62	1.46	1.37	1.24	1.25
29	Outstanding debt principal													
30	Year-end senior lien debt principal		$4,800,000	$10,254,836	$17,133,054	$25,377,438	$34,127,905	$44,931,051	$57,711,443	$70,775,361	$80,072,788	$88,872,175	$124,482,038	$143,326,793
31	Total outstanding debt		$4,800,000	$10,254,836	$17,133,054	$25,377,438	$34,127,905	$44,931,051	$57,711,443	$70,775,361	$80,072,788	$88,872,175	$124,482,038	$143,326,793

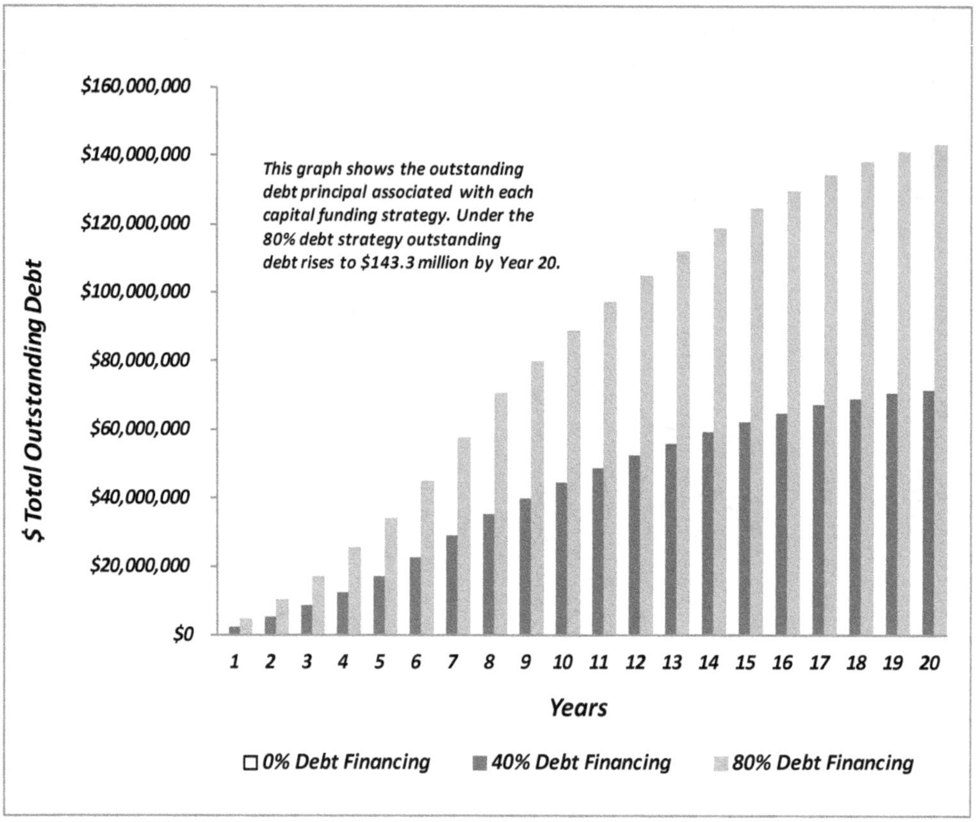

Figure B-1 Outstanding debt principal over 20 years

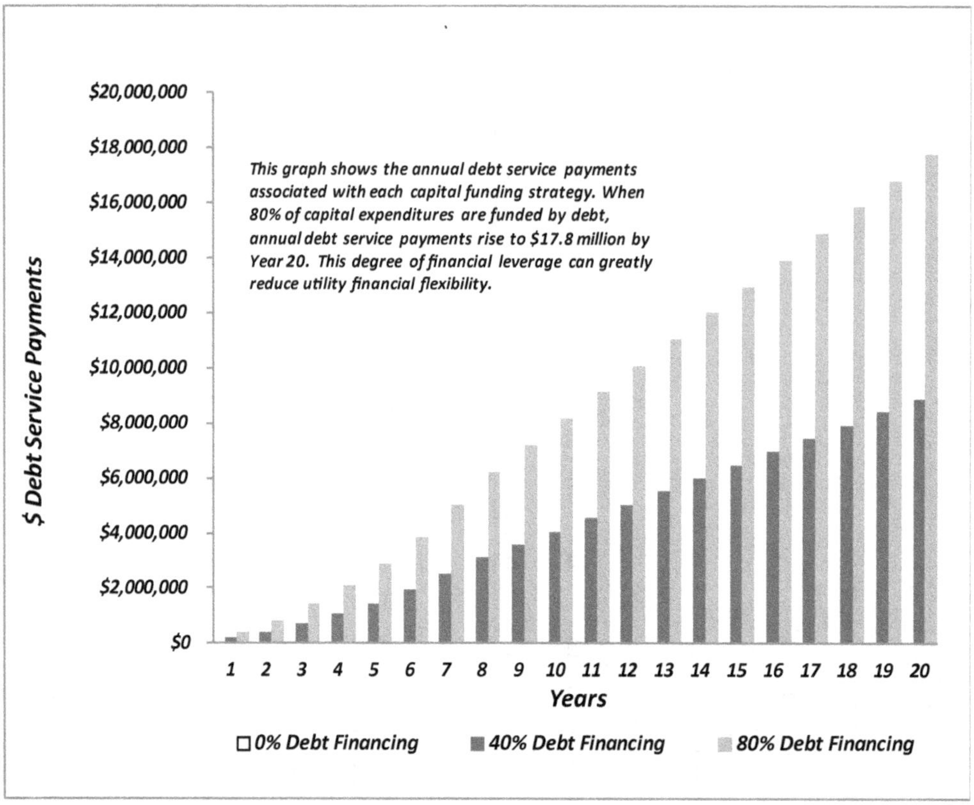

Figure B-2 Outstanding debt service payment over 20 years

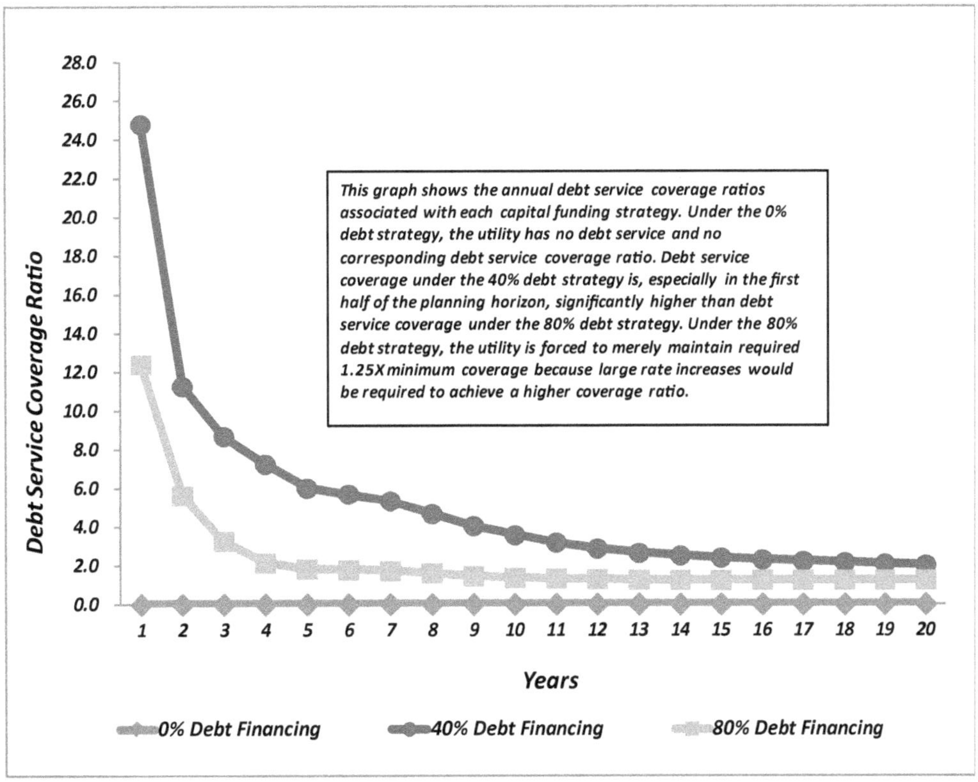

Figure B-3 Debt service coverage ratios

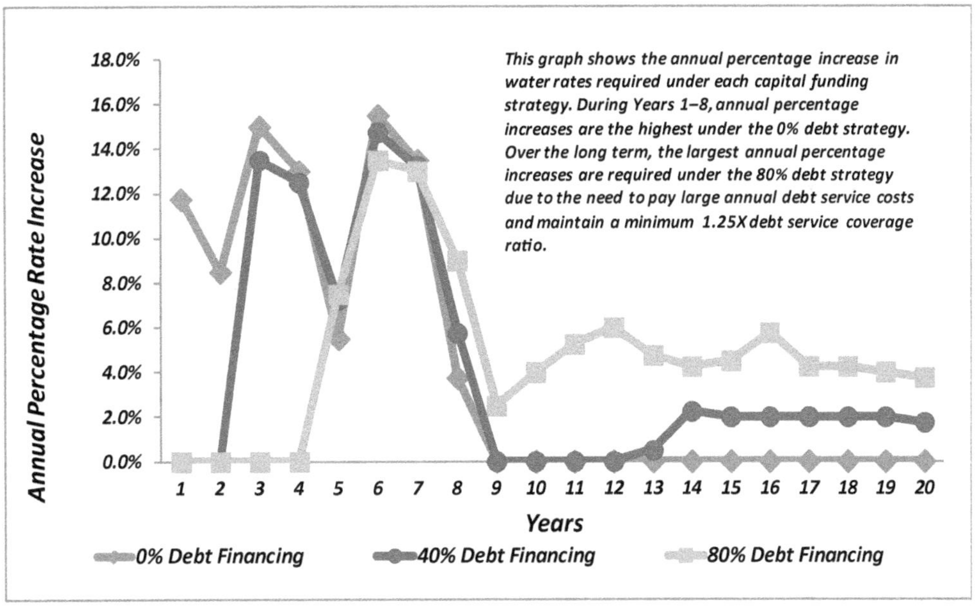

Figure B-4 Annual percentage increases in water rates

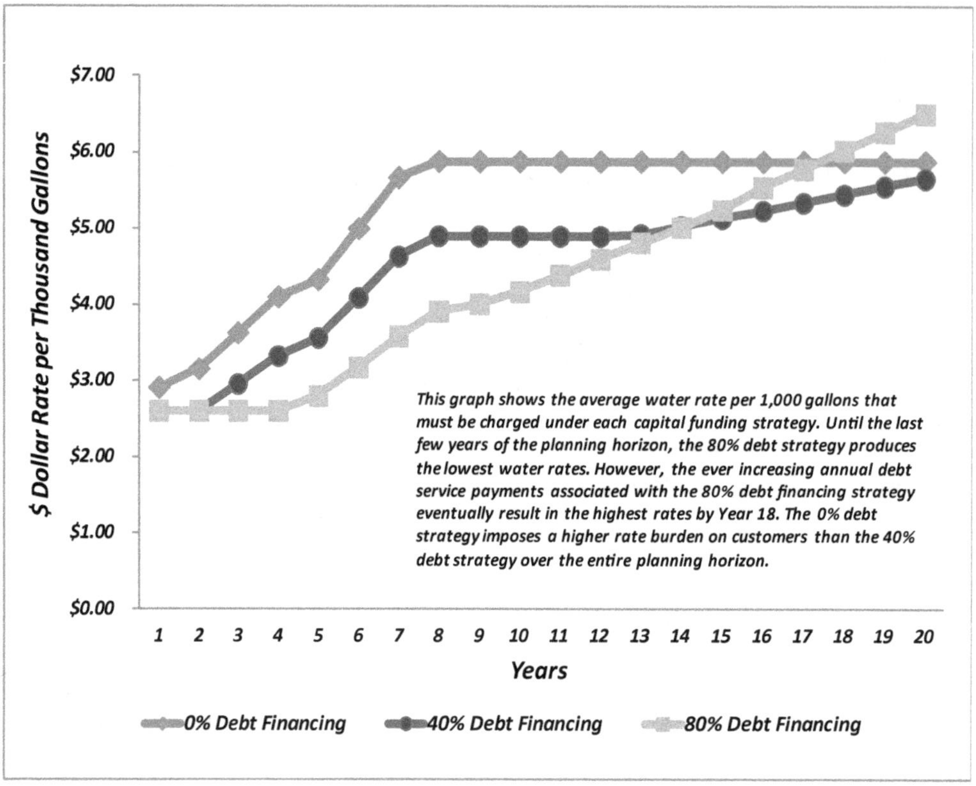

Figure B-5 Average water rate per 1,000 gallons

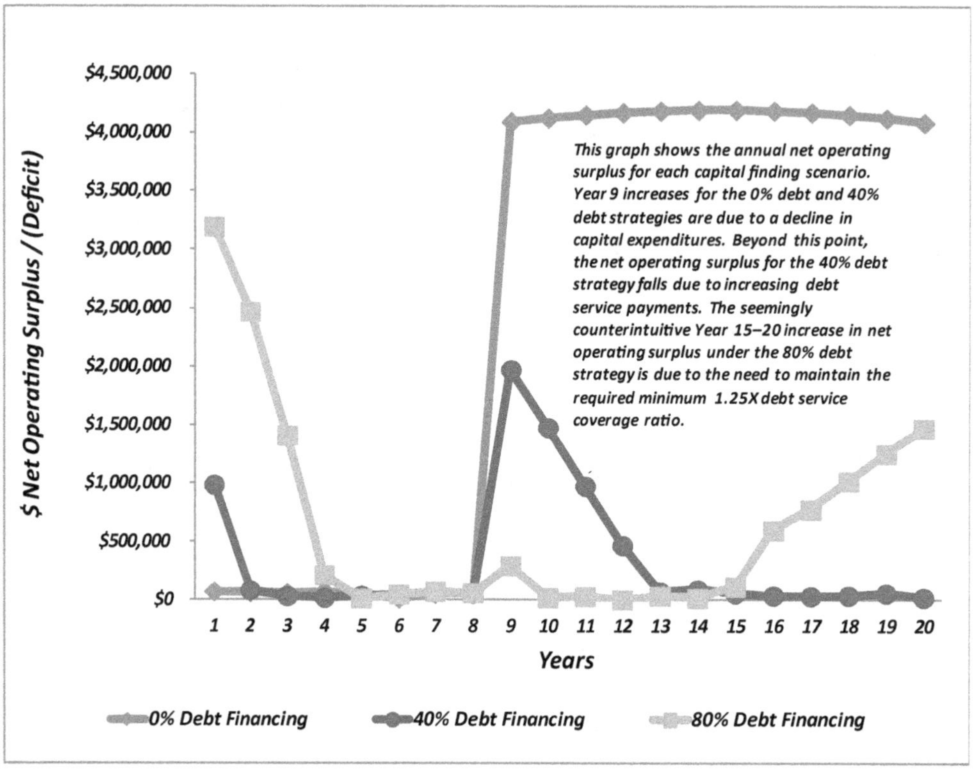

Figure B-6 Annual net operating surplus

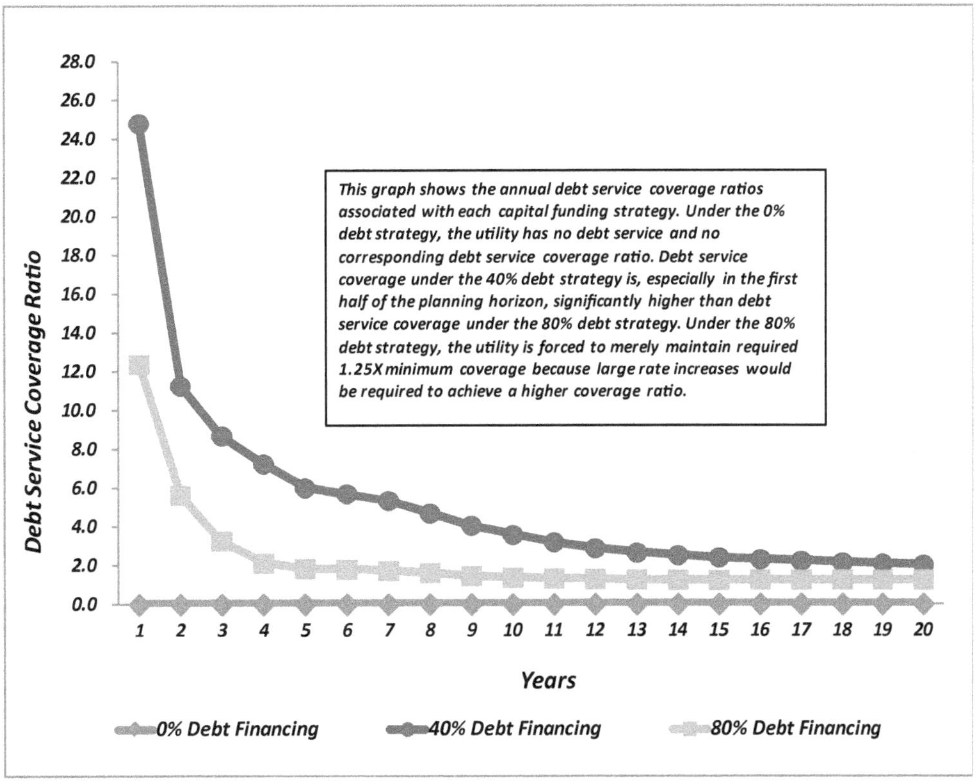

Figure B-7 Operating reserve balance

This page intentionally blank.

Appendix C

Rating Agency Requirements

This appendix discusses the purpose of credit-rating agencies, how rating agencies are compensated, why to obtain a rating, and the rating process. It also examines what a water utility can expect if it decides to request a rating on its debt. Throughout this appendix, the water utility is referred to as the *issuer*. The issuer's financial advisor is frequently a key player in coordinating activities surrounding the rating process (see chapters 4 and 5).

WHAT IS A RATING AGENCY?

Rating agencies are organizations that evaluate and publish opinions with respect to the investment quality of securities (financial obligations). In the United States, nationally recognized statistical rating organizations—Fitch Investors Service, Moody's, and Standard & Poor's—are overseen by the Securities and Exchange Commission. Their ratings have traditionally focused on the creditworthiness (the likelihood of timely repayment of principal and interest) by the issuer with respect to specific financial obligations. Recovery, in case of default, is generally not considered in the rating—only the likelihood a default may occur.

Ratings are based on information supplied by the issuer and information obtained by the rating agency from other sources it considers reliable. Rating agencies do not audit the financial or other information provided to them, and they do not express any opinion as to the appropriateness of any security for any particular investor. All these organizations have adopted policies and procedures established in the Code of Conduct Fundamentals for Credit Rating Agencies issued by the International Organization of Securities Commissions. Although these organizations have no official authority, each has earned the trust of the public through many years of operation in the credit evaluation arena.

Their opinions are generally viewed as independent and objective assessments of the relative creditworthiness of the security being rated.

A rating agency's opinion is expressed in a report that discusses the issuer and the security being rated. The opinion is also summarized symbolically by one or more letters, such as AAA or B. The rating symbol assigned to the security usually appears on the cover of the final official statement and is widely communicated verbally by sellers of the security and in the financial press. Each agency standardizes its opinions so that comparisons can be made of the relative creditworthiness of other securities it has rated. However, while the ratings assigned by different rating agencies are often very similar, they are not identical. Just as people's opinions often differ, so do rating agency opinions. Definitions of long-term debt ratings used by the three organizations appear in Tables C-1, C-2, and C-3.

Different ratings may be caused by different assessments of risks (disclosed or undisclosed contingencies that could adversely affect the issuer's liquidity), the placement of more or less emphasis on various factors, or different opinions about the future impact of probable events. A financial advisor should be aware of and be able to explain any differences between the philosophies and analytical processes of the major agencies.

Credit ratings generally are grouped into two broad categories: investment grade and noninvestment grade. Neither category is meant to indicate which securities are worthy of investment. Only the investor's particular risk preference can determine where that investor will put their money. Many investors, such as state and local governments, are permitted to purchase only investment-grade securities.

In addition to the ratings shown, an agency may place a plus (+) or minus (–) sign or a number (e.g., 1 or 2) with the rating symbol to further distinguish the relative position of a credit within the rating category. In addition, the agency usually assigns an outlook to the rating—positive, negative, stable, or developing; the outlook provides further insight as to any potential or expected changes in the rating. Some agencies may also indicate that the rating is conditional. A conditional rating is generally used when the opinion of credit quality is based on a repayment source that depends on the completion of some act or the fulfillment of some conditions—for example, bonds secured by earnings or rentals from a project under construction. Agencies may, from time to time, indicate that they are reviewing a rating; they may also indicate whether they anticipate that a rating may be raised or lowered as a result of the review. If an issuer fails to provide sufficient information, ratings may be suspended or withdrawn. Because all ratings are security specific, ratings are withdrawn when an issue matures, is called, or is refinanced.

RATING ECONOMICS

Rating agencies are for-profit organizations generating fees through issuers and investors. They rate securities, generally at the request of the issuer, for the benefit of the issuer and the investment community. Some agencies may occasionally issue a rating without having been requested to do so. While this practice may be controversial, agencies will say they only issue an opinion when there is adequate information available to form a credible opinion.

Full rating reports are available to investors by subscription. Rating agencies compete for subscriptions on the basis of quality of analysis, access to information, and price. Because investors may not have either the time or the in-house capability to perform a thorough credit analysis, they frequently base their investment decisions on rating agency evaluations. Institutional investors are increasingly utilizing in-house analysts as well as relying on rating agencies. Rating agencies are often organized by specialty or industry, with designated groups of analysts covering such areas as public finance, industrials, financial companies, and sovereign credits. Within this structure, there will probably be

Table C-1 Fitch Investors Services' rating definitions

	Rating	Description
Investment Grade	AAA	Highest credit quality. AAA ratings denote the lowest expectation of default risk. They are assigned only in cases of exceptionally strong capacity for payment of financial commitments. This capacity is highly unlikely to be adversely affected by foreseeable events.
	AA	Very high credit quality. AA ratings denote expectations of very low default risk. They indicate very strong capacity for payment of financial commitments. This capacity is not significantly vulnerable to foreseeable events.
	A	High credit quality. A ratings denote expectations of low default risk. The capacity for payment of financial commitments is considered strong. This capacity may, nevertheless, be more vulnerable to adverse business or economic conditions than is the case for higher ratings.
	BBB	Good credit quality. BBB ratings indicate that expectations of default risk are currently low. The capacity for payment of financial commitments is considered adequate but adverse business or economic conditions are more likely to impair this capacity.
Noninvestment Grade	BB	Speculative. BB ratings indicate an elevated vulnerability to default risk, particularly in the event of adverse changes in business or economic conditions over time; however, business or financial flexibility exists that supports the servicing of financial commitments.
	B	Highly speculative. B ratings indicate that material default risk is present, but a limited margin of safety remains. Financial commitments are currently being met; however, capacity for continued payment is vulnerable to deterioration in the business and economic environment.
	CCC	Substantial credit risk. Default is a real possibility.
	CC	Very high levels of credit risk. Default of some kind appears probable.
	C	Exceptionally high levels of credit risk. Default is imminent or inevitable, or the issuer is in standstill. Conditions that are indicative of a 'C' category rating for an issuer include: a. The issuer has entered into a grace or cure period following nonpayment of a material financial obligation; b. The issuer has entered into a temporary negotiated waiver or standstill agreement following a payment default on a material financial obligation; or c. Fitch Ratings otherwise believes a condition of 'RD' or 'D' to be imminent or inevitable, including through the formal announcement of a distressed debt exchange.
	RD	Restricted default. RD ratings indicate an issuer that in Fitch Ratings' opinion has experienced an uncured payment default on a bond, loan, or other material financial obligation but which has not entered into bankruptcy filings, administration, receivership, liquidation, or other formal winding-up procedure, and which has not otherwise ceased operating. This would include: a. The selective payment default on a specific class or currency of debt; b. The uncured expiry of any applicable grace period, cure period, or default forbearance period following a payment default on a bank loan, capital markets security, or other material financial obligation; c. The extension of multiple waivers or forbearance periods upon a payment default on one or more material financial obligations, either in series or in parallel; or d. Execution of a distressed debt exchange on one or more material financial obligations.
	D	Default. D ratings indicate an issuer that in Fitch Ratings' opinion has entered into bankruptcy filings, administration, receivership, liquidation, or other formal winding-up procedure, or which has otherwise ceased business.

Note: The modifiers plus (+) or minus (–) may be appended to a rating to show relative standings within the major ratings categories. Such suffixes are not added to the AAA Long-Term Rating category, or categories below B.

Source: Fitch Investor Services. www.fitchratings.com/web_content/ratings/fitch_ratings_definitions_and_scales.pdf

Table C-2 Moody's Investors Service's rating definitions

Grade	Rating	Description
Investment Grade	Aaa	Obligations rated Aaa are judged to be of the highest quality, subject to the lowest level of credit risk.
	Aa	Obligations rated Aa are judged to be of high quality and are subject to very low credit risk.
	A	Obligations rated A are judged to be upper-medium grade and are subject to low credit risk.
	Baa	Obligations rated Baa are judged to be medium-grade and subject to moderate credit risk and as such may possess certain speculative characteristics.
Noninvestment Grade	Ba	Obligations rated Ba are judged to be speculative and are subject to substantial credit risk.
	B	Obligations rated B are considered speculative and are subject to high credit risk.
	Caa	Obligations rated Caa are judged to be speculative of poor standing and are subject to very high credit risk.
	Ca	Obligations rated Ca are highly speculative and are likely in, or very near, default, with some prospect of recovery of principal and interest.
	C	Obligations rated C are the lowest rated and are typically in default, with little prospect for recovery of principal or interest.

Note: Moody's appends numerical modifiers 1, 2, and 3 to each generic rating classification from Aa through Caa. The modifier 1 indicates that the obligation ranks in the higher end of its generic rating category; the modifier 2 indicates a mid-range ranking; and the modifier 3 indicates a ranking in the lower end of that generic rating category.

Source: Moody's Investors Services Global Long-Term Rating Scale. www.moodys.com/researchdocumentcontentpage.aspx?docid=PBC_79004

Table C-3 Standard & Poor's rating definitions

Grade	Rating	Description
Investment Grade	AAA	Extremely strong capacity to meet financial commitments. Highest Rating.
	AA	Very strong capacity to meet financial commitments.
	A	Strong capacity to meet financial commitments, but somewhat susceptible to adverse economic conditions and changes in circumstances.
	BBB	Adequate capacity to meet financial commitments, but more subject to adverse economic conditions.
	BBB–	Considered lowest investment grade by market participants.
Noninvestment Grade	BB+	Considered highest speculative grade by market participants.
	BB	Less vulnerable in the near-term but faces major ongoing uncertainties to adverse business, financial, and economic conditions.
	B	More vulnerable to adverse business, financial, and economic conditions but currently has the capacity to meet financial commitments.
	CCC	Currently vulnerable and dependent on favorable business, financial, and economic conditions to meet financial commitments.
	CC	Currently highly vulnerable.
	C	Currently highly vulnerable obligations and other defined circumstances.
	D	Payment default on financial commitments.

Note: Ratings from 'AA' to 'CCC' may be modified by the addition of a plus (+) or minus (–) sign to show relative standing within the major rating categories.

Source: Standard & Poor's Financial Services. www.standardandpoors.com/ratings/definitions-and-faqs/en/us

another level of specialization. It may be by type of issuer (e.g., airports, utilities, housing, banking) or by geographic region. Because the quality of the analysis varies with the skill and experience of the rating agent assigned, investors often have a preference for a certain organization's ratings in a particular specialty area. Most investors do not subscribe to all the rating agency services.

When it requests a rating, an issuer is charged a fee by the agency and may also be assessed an annual fee while the securities are outstanding. The issuer must evaluate the economics of whether to obtain a rating based on the anticipated cost of financing with and without a rating. Issuers seek credit ratings to improve the marketability or pricing of the financial obligation. Because many investors buy only rated securities, having a rating broadens the potential pool of buyers. A broader base of buyers should increase demand for the security, thereby lowering the overall cost of financing. For issues of a very small dollar amount, however, the cost of the rating may be greater than the interest savings, especially if there are local investors who are familiar with the issuer.

In deciding whether to get one or more ratings and in choosing which agencies to engage, the issuer often looks at two factors. One factor is which rating agencies are most frequently used by probable buyers of the security. The other factor is the probability of getting the desired rating from a particular agency given the differences in the credit factors emphasized by the various agencies.

THE PROCESS

The issuer must apply to the agency for a rating and must provide information the agency needs to form an opinion about the credit quality of the debt to be sold. The rating agency will also obtain information from other sources it considers reliable.

Documentation

To answer the questions that arise when evaluating the security, the rating agency should review various types of information, both financial and nonfinancial. The most commonly requested documents are listed in Table C-4. Some of the documents are provided only the first time an issuer obtains a rating; others are provided or updated annually. Still others are only available a few weeks before the sale. The issuer's financial advisor should help determine which documents are relevant and when they will be needed. Some advisors also provide assistance with gathering information.

The rating request is entered on a calendar and assigned a priority based on the anticipated sale date. The application is then assigned to an analyst. The analyst reviews the material submitted, becomes familiar with the issuer and the security structure, calculates various financial ratios, and may, if necessary or helpful, meet with the issuer to clarify information.

The analyst's evaluation and rating recommendation are condensed into a report presented to the analyst's manager for review. The analyst and manager then present their conclusions to a rating committee. The rating committee's role is to assign a rating that reflects the creditworthiness of the security relative to all other securities rated by the agency. This requires familiarity with issues and concerns that cross industries.

As part of the rating process, the issuer generally agrees to provide continuing information to the rating agency. Based on its evaluation of the information provided and other generally available information, the rating agency may from time to time confirm or change the rating on outstanding securities. Changes are communicated to subscribers and are often reported in the financial press.

Table C-4 Rating agencies' frequently requested documents

Information Category	Documentation Requested
General	Description of the security
	Preliminary official statement
	Bond ordinance, resolution, indenture, plus supplemental documents
	Bond counsel opinion
	Flow of funds
	Utility description and capacity (historic and projected)
	System map
	Organizational structure
	Description of facilities
	Average and peak demands and water loss
	Redundancy and security
	Growth of service area
	Customer information
	Service contracts with wholesale customers
	Largest customers by revenue and consumption
	Revenue and consumption by customer class
	Service area economics and demographics
	Population/income trends
	Employment data/unemployment rates
	Tax base trends/building permits
	Largest employers
	Labor relations history and labor contract summaries
	Regulatory compliance status and strategies
	Engineering report, feasibility study, or rate study
Financial	Capital improvement program
	Anticipated capital and financial plans or pro forma projections
	Water rates and comparison to rates in surrounding communities
	History of rate increases
	Accounts receivable/collections
	Current year budget/budget process
	Audited financial reports for 3 to 5 years
	Financial ratios

Relevant Questions

When evaluating a debt issue, the rating agency is most concerned with the long-term likelihood of timely repayment of principal and payment of interest. It will ask the same questions any lender would consider. These key questions are also referred to as the three Cs of credit: character, capacity, and collateral. Detailed questions are listed in Table C-5.

Table C-5 Three Cs of credit applied to the water industry

Category	Relevant Questions
Character: Willingness to Pay	Is this a reputable issuer? Does it have history of meeting its obligations when due? Is its management reliable? What is the purpose of the borrowing? Is the project(s) being financed essential or nonessential to the operation of the organization?
Capacity: Ability to Pay	Will cash flows from the project or net operating revenues of the organization be sufficient to service the debt? Are the net operating revenues legally available to service the debt? Can additional debt be issued? If so, are there any limitations on the amount or any preference for holders of this issue? How will additional debt affect the organization's ability to make timely payment of principal and interest on this security? What intergovernmental relationships are in place? How reliant are the various government units on intergovernmental transfers? Is management competent? Do key personnel have the experience needed to operate the organization in an efficient manner? How long have the key players been in their positions? Is the organization stable or in chaos? Is past performance likely to be representative of future performance? Is the water supply adequate? What population growth is projected in the service area? What is the condition of the physical plant? Are there any regulatory compliance issues that will require major capital expenditures? How large is the capital improvement plan during the next 5 to 10 years? How does the organization plan to finance its capital investments? What rate structure is in place? What is the rate-setting philosophy and process? How easy is it to obtain rate increases? Who has control and who can make decisions? What is the organization's relationship with any rate regulatory authorities? What are the service area demographics? What is happening with employment and income in the service area? Can the population afford rate increases?
Collateral: Security	Have revenues been pledged to this issue? Is there a commitment to raise rates as necessary to service the debt? If revenues are insufficient, are property taxes available for debt service? What constitutes a default? What are the bond covenants if there is a mortgage lien? What legal protections do security holders have? What commitments have been made with respect to disclosure of the future information that may affect the security?

The committee's rating opinion is released first to the issuer and then, at the discretion of the issuer, to the public. The release is followed shortly by distribution of the full credit report. The issuer generally has an opportunity to review the report before its release. Because the issuer is paying for the rating, there may be times that the issuer does not wish the report to be released to the public—for example, the issuer only chooses to go with one rating. The assignment of the rating needs to be closely coordinated with the issuer's calendar, especially in the case of a competitive sale.

COMMUNICATING WITH THE RATING AGENCY

It is essential that a small, first-time, or infrequent issuer discuss the rating process with its financial advisor. The financial advisor should be able to provide insight into which agencies would be the best fit for the issuer and to help assess what rating will probably be assigned and under what circumstances. The financial advisor may already know the industry analyst at each of several rating agencies. Although it is common for the financial advisor to make initial contacts with appropriate analysts, a large, frequent issuer may feel comfortable performing these tasks in-house.

Communication with the rating agency takes multiple forms. The first is through documents that the issuer provides to the agency. The second is via an informal question-and-answer process as the analyst reviews the documents. The financial advisor handles a great deal of this communication. It is essential for the advisor to know the organization well or at least to have access to the appropriate issuer staff to obtain timely and accurate clarification of points arising during the document review process. The more unusual or complicated a security, the more time the rating agency will need to review materials, and the more likely it is that questions will arise during the review. More time and clarification will also be required to review materials from first-time or infrequent issuers than from an issuer with whom the analyst is familiar.

A third form of communication with the rating agency is the formal meeting. Meetings give the analyst the opportunity to assess management's competency, philosophy, and character. A meeting also gives management a chance to present its history and to explain complex issues and strategies. Top-level financial and operating managers of the issuer should be involved in the presentation, as should the financial advisor. The meeting can be structured as a phone conference, or it can be an in-person visit held either at the rating agency's office or at the issuer's facility. A meeting at the issuer's facility has several advantages:

- More of the issuer's staff can be involved—it is important to convince the analyst that the organization is well run and that competency and depth exists throughout the organization.
- A facility tour can be used to demonstrate good operating practices, show assets previous debt issues have funded, new or unique treatment processes, etc.
- Local officials who may be unable to commit the time to travel to out-of-town meetings may be able to participate for a portion of the meeting.

A site visit does require more of the analyst's time, however, so it may not be reasonable to expect every rating to involve an in-person meeting at the issuer's facility. In preparing for a meeting, the issuer should focus on presenting its best side, but it must also be open and forthright in addressing any potential issues or concerns the analyst may have. Because one of the purposes of the meeting is to give the analyst an opportunity to assess the character and integrity of management, trying to hide problems is counterproductive. The meeting will often be structured with an issuer presentation followed by a period during which the analyst asks questions. If the answer to a question is not known, advise the analyst that the answer will be found. One person (usually the financial advisor) in the meeting should be assigned to documenting any issues requiring additional information or clarification. Color charts, graphs, booklets, and summaries that the analyst can retain from the meeting are often useful. Be prepared to talk about the following issues:

- Overview of the utility
- Major issues currently facing the utility and those for the next five years
- Economic and demographic conditions in the service area (stability, population trends, customer profile and usage trends, top customers, etc.)
- Current operating statistics (system capacity, average day, maximum day, minimum day, water loss, water rights, aquifer depletion, etc.)
- Capital improvement plan (what will the funds received be used for?)

- The organization's audited financial statements (past three to five years)
- Water rates (compared to neighboring utilities, compared to similar systems, customer affordability, rate-setting process, etc.)
- Pro forma financial projections (usually five years), assumptions, and future strategies
- Legal provisions (rate covenants, coverage levels, security, additional bonds tests, flow of funds, reserve funds, etc.)
- Plan of finance (when will the rating be needed, date of sale, etc.)

It is also helpful for the issuer and the rating agency representatives to agree on a meeting agenda in advance. The agency representatives will want to be sure that adequate time will be allotted for their questions during or after the issuer's presentation so that no essential information will be missed because of time constraints.

An effective communication strategy will involve continued contact after a rating has been issued. Future ratings applications will be less traumatic if the analyst is familiar with the organization and has respect for its management. Material changes in the organization should be communicated to the analyst between scheduled rating visits, and the rating agency will also want to receive regular financial reports. Some analysts also like to receive the organization's news releases and other public communications. An issuer should discuss with the financial advisor and the rating agency analyst the preferred level of communication between rating meetings.

ADDITIONAL INFORMATION

Each of the rating agencies has materials prepared to help issuers and investors understand the role of the rating agency, the agency's rating systems, and its financial analysis process. Those materials can be obtained directly from the rating agency, their websites, or they should be available from the issuer's financial advisor. Other organizations, such as the Government Finance Officers Association and the Treasury Management Association, have also published material that may be useful to those who need to know more about the rating process. A listing of current publications can be obtained directly from those organizations. Many of the larger underwriters and financial advisory firms also publish booklets that contain useful information.

SUMMARY

A water utility, like other organizations that issue debt, may want to have a major credit rating agency publish an opinion about the credit quality of its debt and enhance the marketability of the issue. Investors often base their investment decisions on a rating agency's opinion regarding the likelihood an issuer will repay its financial obligation. A utility that decides to obtain a rating will also need to decide, possibly with the assistance of its financial advisor, which agency is most appropriate for the organization and whether to obtain more than one rating.

The utility should be prepared to provide information to the rating agency about the security to be sold. In addition, the rating agency will want to receive information about the utility including its legal structure, its financial status, its operations, and the economic environment in which it operates. After the security is sold, the utility is expected to provide certain updated information as long as the security remains outstanding.

This page intentionally blank.

Glossary

accelerated depreciation Use of depreciation methods that amortize the cost of an asset at a faster rate than under the straight-line method. The three principal methods of accelerated depreciation are (1) sum of the year's digits, (2) double declining balance, and (3) units of production.

accountant A financial professional who specializes in accounting. In connection with a bond issue, an accountant often provides the most recent audited financial statements and may provide escrow verification.

accrual basis A basis of accounting in which revenues are recognized when earned and expenses are recognized when they become liabilities for benefits received. Receipt of revenue or payment of the expenditure may take place, in whole or in part, in another accounting period.

additional bonds test A requirement in an indenture that additional bonds not be issued unless historical and projected revenues indicate there is sufficient revenue to avoid dilution of coverage on outstanding bonds

ad valorem tax A state or local tax based on the assessed value of real or personal property

advance for construction An advance that may be refundable either wholly or in part and that is made to the utility in order to fund construction. When all potential claims for refunds have been settled, the balance, if any, is treated as a contribution-in-aid-of-construction, under generally accepted accounting principles.

advance refunding bonds Bonds issued to replace an outstanding bond issue before the date on which the outstanding bonds become due or callable. Proceeds of the advance refunding bonds are deposited in escrow with a fiduciary, invested in US Treasury bonds or other authorized securities, and used to redeem the underlying bonds at maturity or on the call date and to pay interest on the bonds being refunded or on the advance refunding bonds.

AFUDC allowance for funds used during construction

all-in debt-service coverage Annual revenues available for debt service divided by the total annual debt service

amortization The distribution of debt payments into multiple cash-flow installments

ancillary charge A separate charge for ancillary services that is not included in costs for general water service. In providing water service, a utility must often perform these types of services, which often benefit only the individual customer using the services and have no system-wide benefit. Such services might include: account activation, connect and disconnect activities, meter testing, etc.

asset management plan A plan developed to manage the assets of a utility system

authority bonds Bonds payable from the revenues of a specific authority, such as a water or sewer authority. Because authorities usually have no revenues other than charges for services, their bonds are ordinarily revenue bonds.

availability charge A limited-use charge made by a water utility to a property owner between the time when water service is made available to the property and the time when the property connects to the utility's facilities and starts using the service

BAN bond-anticipation notes: short-term interest-bearing notes issued in advance of a longer-term future bond issue

betterment An addition or change that is made to a fixed asset and is expected to prolong that asset's life or to increase its efficiency to an extent greater than normal maintenance would accomplish. The cost of the betterment is added to the book value of the asset.

bond A written promise to pay a specified sum of money, called the face value or principal amount, at a specified date or dates in the future, called the maturity date(s), together with periodic interest at a specified rate

bond counsel An attorney who provides expertise in securities law

bond discount The amount of the face value of a bond in excess of the price for which the bond is acquired or sold. The price does not include accrued interest at the date of acquisition or sale.

bond indenture A document that specifies the legal obligation of the bond issuer. Typically, such an instrument contains pledges to bondholders regarding payment of principal and interest, operation, flow of funds, further debt issuance, what will constitute a default, and remedy available in the event of default.

bond ordinance *See* bond indenture

bond premium The amount by which a bond is acquired or sold in excess of its face value. The price does not include accrued interest at the date of acquisition or sale.

bond purchase agreement A legally binding document between a bond issuer and an underwriter that establishes the terms of the bond sale

bond resolution *See* bond indenture

bonded debt That portion of indebtedness represented by outstanding bonds. *See also* net bonded debt.

BPA bond purchase agreement

CAB capital appreciation bond

CAC customer-advance-for-construction

call (1) Redeeming a bond or preferred stock issue before its normal maturity. (2) An option to buy (or call) a share of stock at a specified price within a specified period.

capital appreciation bond (CAB) A bond that pays no interest on a periodic basis, but accretes in value from the date of issuance to the date of maturity

capital budget A plan that covers all major additions, replacements, and repairs greater than a maintenance expenditure to the existing utility plant

capital improvement program A plan used to coordinate the type, location, timing, and financing of capital improvements over a multiyear period

certificate of participation (COP) A security evidencing an interest in rental payments made according to a lease arrangement. In most cases the lessee (the local government) determines the lease provisions subject only to market acceptance and rating agency approval because the lessor is an entity created by the lessee to sell the securities.

CIAC contributions-in-aid-of-construction

CIP capital improvement program

closed-loop system A system whereby the revenues generated by the system stay within the system to pay for current or future system expenses

commercial bank A financial institution that accepts deposits and offers loans. Commercial banks may offer tax-exempt loans or credit facilities to traditional municipal issuers.

commercial paper (CP) An unsecured short-term promissory note with an average maturity of 30 to 45 days. The notes have a maximum maturity of 270 days and are intended to be refinanced (rolled over) continuously for periods that may exceed one year.

competitive sale A form of bond sale in which the issuer publishes a notice of sale that specifies, among other things, when bids are due and how they will be evaluated.

connection charge A fee assessed by a utility to recover the cost of connecting a customer's service line to the utility's facilities

construction-work-in-progress (CWIP) A utility's investment in facilities that are under construction but are not yet dedicated to service. Until that property is placed in service, some regulatory agencies exclude this investment from the rate base on which utilities are permitted to earn a return.

consulting engineer A professional engineer who is engaged for a specific purpose. With respect to a debt issue, a consulting engineer generally reviews the operations and capital plans and provides a report that is included in the official statement.

contributions-in-aid-of-construction (CIAC) An amount of money, services, or property that is received by a water utility from any person, governmental agency, or other entity and that is provided at no cost to the utility. It represents an addition or transfer to the capital of the utility and is used to offset the acquisition, improvement, or construction costs of the utility's property, facilities, or equipment used in providing utility services to the public. It includes amounts transferred from advances for construction representing any unrefunded balances of expired refund contracts or discounts resulting from termination of refund contracts. Contributions received from governmental agencies and others for relocation of water mains or other plant facilities are also included.

COP certificate of participation

cost of service The total annual operation and maintenance expense and capital-related costs incurred in meeting various aspects of providing water utility service

coverage ratio Annual net revenues divided by annual debt service

CP commercial paper

CPA certified public accountant

CWIP construction-work-in-progress

credit enhancer An entity that insures bonds or provides a letter of credit to back bonds

customer-advance-for-construction (CAC) Payments received from customers for installation of facilities to serve new developments

debenture A type of debt that is not secured by physical assets or collateral. Rather, they are backed only by the general creditworthiness and reputation of the issuer.

debt ceiling An upper limit on the amount that can be borrowed

debt portfolio A grouping, or the total amount, of debt held by a utility

debt service The cash that is required to pay the principal and interest amounts on debt over a period of time

debt-service reserve fund A fund established to cover debt service for some period of time in the event of a revenue shortage, often one-year's debt service

defease To refund the bonds such that the legal documents that controlled the security provisions for the bonds no longer apply

deferred taxes Taxes that are charged against current operating results or reflected in the income statement and that are not currently payable to the government. Cash generated by deferring taxes is a source of funding for capital improvements.

depreciation An indication of the amount of service value not restored by current maintenance of depreciable utility plant facilities. Among the causes of this loss in value are wear and tear, decay, action of the elements, inadequacy, obsolescence, changes in demand, and requirements of public authorities.

disclosure counsel An attorney retained by the issuer to conduct due diligence, draft the official statement, and provide guidance with respect to ongoing disclosure responsibilities

double-barrel bonds Revenue bonds backed by a specific source of revenues to which the government adds a general obligation pledge. These bonds are sometimes called combination bonds.

EMMA Electronic Municipal Market Access

financial feasibility consultant A financial professional engaged to issue a report with respect to the financial feasibility of either the utility as a whole or a specific project

floating-rate demand note Securities on which the interest rate changes at predetermined intervals (often monthly) and that give the purchaser of the security the right to demand that the seller (utility) redeem the notes at each interest adjustment date

financial printer A printer that specializes in the printing of official statements, notices of sale, and other items of a financial nature

GAN grant-anticipation note

GASB Governmental Accounting Standard Board

general obligation (GO) bonds A debt obligation issued by a government entity, such as a state or municipality, and backed by a pledge of the entity's full faith and credit, including its taxing authority

GO bond general obligation bond

governmental utility *See* publicly owned utility

grant-anticipation note (GAN) Short-term interest-bearing notes issued in advance of the receipt of a grant

indenture A legal and binding contract between a bond issuer and the bondholders

integrated resource planning (IRP) A process that maximizes available resources by considering a wide range of supply-side and demand-side resources

investor-owned water utility A utility owned by an individual, partnership, corporation, or other qualified entity, with the equity provided by shareholders. Investor-owned utilities are generally subject to some form of regulation.

IOSCO International Organization of Securities Commissions

IRP integrated resource planning

issue The process of offering securities as an attempt to raise funds

issuer A borrower; in this manual, generally the water utility. This entity determines the need for bonding and selects the financing team.

issuer's counsel An attorney who represents the issuer and who reviews local legal matters for inclusion in the official statement

investor-owned utility Systems where the physical assets are owned by an individual or group of individuals. Such utilities employ an equity-based ratemaking approach with the revenue requirements aligning more with financial reporting under generally accepted accounting principles. Also called private utilities.

lease An agreement in which the owner of property (lessor) permits another party (lessee) to use the property in exchange for an agreed-on payment

letter of credit (LOC) An agreement by a bank to make a payment to the beneficiary if certain documents are presented to the bank. When the agreement obligates the bank to make principal and interest payments on bonds if the issuer is in default, a letter of credit can allow the issuer to obtain a higher credit rating.

leveraged lease A lease in which the lessor borrows the funds to acquire the property under lease and the lease payments are used to retire the debt

lien The legal right of a creditor to sell the collateral property of a debtor who fails to meet the obligations of a loan contract

line of credit An arrangement between a lender, usually a bank, and a borrower to make funds available as needed

LOC letter of credit

long-term debt For financial statement purposes, debt having a maturity in excess of 1 year. Debt with a maturity of 2 to 10 years is often called medium-term debt.

MADS maximum annual debt service

maximum annual debt service (MADS) The maximum amount of principal and interest due by a revenue bond issuer on its outstanding bonds in any future fiscal year

medium-term note (MTN) A debt note that usually matures (is paid back) in 5 to 10 years

MHI median household income

MTN medium-term note

NAIC National Association of Insurance Companies

NARUC National Association of Regulatory Commissioners

MSRB Municipal Securities Rulemaking Board

negotiated sale A bond sale in which the issuer negotiates a price directly with the underwriter rather than selling the bonds on the market

net bonded debt Bonded debt less any cash or other assets that are available and earmarked for that debt's retirement

NRMSIR Nationally Recognized Municipal Securities Information Repository

NRSRO Nationally Recognized Statistical Rating Organizations

note A short-term debt security, usually with a maturity of five years or less

official statement (OS) A document that contains all information that a potential investor is expected to need about a security and the issuer. The preliminary official statement is issued before the sale; the final official statement is issued within 10 days after the sale and includes the prices at which the securities were offered to the public.

OID bond original-issue discount bond

OS official statement

original-issue discount (OID) bonds Long-term bonds sold at a price below par because they have an interest rate substantially lower than prevailing market rates

paying agent An entity, often a bank trust department, that is engaged by an issuer to disperse principal and interest to bondholders

par The face value of a stock, bond, or other security

POS preliminary official statement

preliminary official statement (POS) A document issued before the sale that contains all information that a potential investor is expected to need about a security and the issuer, except the prices. *See* official statement.

private placement A method of sale in which the issuer sells bonds directly to a small number of sophisticated investors

publicly owned utility A water utility created by legislative action of a state or other government agency. A government-owned water utility may be part of a municipal government operation, a county agency, or a regional authority, or it may take such other forms as are appropriate for its service area. Also called governmental utilities.

public–private partnership Government service or private business venture that is funded and/or operated through a partnership of government and one or more private sector companies

put The right of an investor to require an issuer to repurchase bonds at a predetermined time and price

put bonds *See* tender-option bonds

RAN revenue-anticipation note

rate base The amount of capital invested by the utility in order to provide utility services. A typical determination of rate base includes utility plant in service, less accumulated depreciation, less accumulated deferred income taxes, less contributions in aid of construction, plus a working capital allowance.

rate covenant A requirement stating that an issuer must pledge in a bond document to set rates sufficient to meet all operating costs and some multiple of debt service

rating agency An independent organization that publishes, in the form of a rating, its opinion about the creditworthiness of a security

rate stabilization fund A mechanism for setting aside money either for unforeseen needs or for capital projects

registrar An organization, often a bank trust department, engaged by the issuer to maintain records of bond ownership

revenue-anticipation note (RAN) Short-term interest bearing notes that are repaid with expected revenues from the project being financed by the notes

revenue bonds A bond secured by and payable exclusively from revenues received from system operations or from the project being financed

revenue pledge An obligation to use specified revenues for debt service

reserves money set aside to meet costs that may arise in the future or for the future costs of upkeep

SCADA supervisory control and data acquisition

SDC system development charge

security A fungible, negotiable financial instrument that represents some type of financial value

SEC Securities and Exchange Commission

S&P Standard & Poor's Corporation

state revolving fund (SRF) A program typically administered by a state department of environmental quality or health to provide loans to water utilities for capital improvements at interest rates at or below market rates. The Safe Drinking Water Act, as amended in 1996, established the Drinking Water State Revolving Fund (DWSRF) to make funds available to drinking water systems to finance infrastructure improvements. The Clean Water State Revolving Fund (CWSRF) program is part of the Clean Water Act and provides low-cost financing for a wide range of water quality infrastructure projects.

system development charge (SDC) A contribution of capital required from new utility customers or existing customers requesting enlarged or expanded services and applied toward existing or planned plant facilities necessary to meet the service needs of those customers. These charges are intended to provide funds to be used to finance all or part of capital improvements necessary to serve new customers. These charges may also be called impact fees, plant investment fees, and capital recovery charges.

subordinated debt Debt that has a lower priority than other bonds of the issuer in case of liquidation during bankruptcy

TAN tax-anticipation note

tax-benefit district A specific geographic area that is designated a unique charge that governments can assess against real estate parcels for certain public projects

tax-anticipation note (TAN) Short-term interest bearing notes that are repaid with future tax collections

tax-exempt commercial paper (TECP) A short-term promissory note with an average maturity of 30 to 45 days. Such notes, which have a maximum maturity of 270 days, are intended to be refinanced (rolled over) continuously for periods that may exceed 1 year.

TECP tax-exempt commercial paper

tender-option bonds Long-term bonds that provide the investor with the option of requiring the issuer to repurchase the bonds, generally at par, at a specified date or dates in advance of the stated maturity. Often called put bonds.

trustee An organization, generally a bank trust department, that holds funds and is responsible for protecting investors' rights

underwriter An entity, generally an investment banking firm, that purchases bonds from an issuer and resells them to investors

underwriter's counsel An attorney who represents the interests of the underwriter and is responsible for document preparation and review. Underwriter's counsel prepares a purchase contract between the issuer and the underwriter, as well as various underwriting documents and securities filings that are of concern only to the underwriter. He/she may also prepare the official statement.

user charges Fees or rates or both that are paid by customers of the system for water service

Water Infrastructure Finance and Innovation Authority (WIFIA) Part of the Federal Water Resources Reform and Development Act that was signed into law in 2014 that provides loan guarantees and direct loans for water- and wastewater-related infrastructure projects at long-term Treasury rates

wrapping A way of structuring debt that enables the utility to maximize its ability to meet annual cash requirements and debt covenants by considering the structure of other existing debt obligations so new obligation payments will have the least impact on the issuer

zero-coupon bonds Long-term bonds that pay no interest before maturity originally sold at a substantial discount from par

This page intentionally blank.

Index

F

G

I

L

M

T

U

V

W

www.ingramcontent.com/pod-product-compliance
Lightning Source LLC
LaVergne TN
LVHW082339200926
848481LV00012B/3388
* 9 7 8 1 6 2 5 7 6 0 1 6 6 *